Introduction to
Dairy Microbiology

Introduction to
Dairy Microbiology

Ranvijay Singh

Editor

KOROS PRESS LIMITED
London, UK

Introduction to Dairy Microbiology

© 2012

Printed in 2017 for Sale in the Indian Subcontinent

Published by
Koros Press Limited
3 The Pines, Rubery B45 9FF, Rednal,
Birmingham, United Kingdom

Tel.: +44-7826-930152
Email: info@korospress.com
www.korospress.com

ISBN: 978-1-78163-182-9

Editor: Ranvijay Singh

10 9 8 7 6 5 4 3 2 1

British Library Cataloguing in Publication Data
A CIP record for this book is available from the British Library

Exclusively distributed by CBS Publishers & Distributors Pvt. Ltd.
Sales & Distribution Rights only for India, Pakistan, Bangladesh, Sri Lanka, Nepal and Bhutan.This book is not to be sold outside these territories.

Contents

Preface

The microbial quality of raw milk is crucial for the production of quality dairy foods. Spoilage is a term used to describe the deterioration of a foods' texture, colour, odour or flavour to the point where it is unappetizing or unsuitable for human consumption. Microbial spoilage of food often involves the degradation of protein, carbohydrates, and fats by the microorganisms or their enzymes. In milk, the microorganisms that are principally involved in spoilage are psychrotrophic organisms. Most psychrotrophs are destroyed by pasteurization temperatures, however, some like *Pseudomonas fluorescens, Pseudomonas fragi* can produce proteolytic and lipolytic extracellular enzymes which are heat stable and capable of causing spoilage. Some species and strains of *Bacillus, Clostridium, Cornebacterium, Arthrobacter, Lactobacillus, Microbacterium, Micrococcus,* and *Streptococcus* can survive pasteurization and grow at refrigeration temperatures which can cause spoilage problems.

Hygienic milk production practices, proper handling and storage of milk, and mandatory pasteurization has decreased the threat of milkborne diseases such as tuberculosis, brucellosis, and typhoid fever. There have been a number of foodborne illnesses resulting from the ingestion of raw milk, or dairy products made with milk that was not properly pasteurized or was poorly handled causing post-processing contamination. Bacteriophages are viruses that require bacteria host cells for growth and reproduction. Initially, the bacteriophage attaches itself to the bacteria cell wall and injects nuclear substance into the cell. Inside the cell, the nuclear substance produces shells, or phage coats, for the new bacteriophage which are quickly filled with nucleic acid. The bacterial cell ruptures and dies as the new bacteriophage are released.

Bacteriophages are ubiquitous but generally enter the milk processing plant with the farm milk. They can be inactivated heat treatments of 30 min at 63 to 88° C, or by the use of chemical disinfectants. Bacteriophages are of most concern in cheese making.

They attack and destroy most of the lactic acid bacteria which prevents normal ripening known as slow or dead vat. Commercial manufacturers provide starter cultures in lyophilized (freeze-dryed), frozen or spray-dried forms. The dairy product manufacturers need to inoculate the culture into milk or other suitable substrate.

The book has been written keeping in mind for the graduate and post graduate level bio-sciences and interdisciplinary courses.

—Ranvijay Singh

Chapter 1

Microbiology of the Cream

Cream is a dairy product that is composed of the higher-butterfat layer skimmed from the top of milk before homogenization. In unhomogenized milk, over time, the lighter fat rises to the top. In the industrial production of cream this process is accelerated by using centrifuges called "separators". In many countries, cream is sold in several grades depending on the total butterfat content.

Cream can be dried to a powder for shipment to distant markets. Cream skimmed from milk may be called "sweet cream" to distinguish it from whey cream skimmed from whey, a by-product of cheese-making. Whey cream has a lower fat content and tastes more salty, tangy and "cheesy".

Cream produced by cows (particularly Jersey cattle) grazing on natural pasture often contains some natural carotenoid pigments derived from the plants they eat; this gives the cream a slight yellow tone, hence the name of the yellowish-white colour, cream. Cream from goat's milk, or from cows fed indoors on grain or grain-based pellets, is white.

Types

Different grades of cream are distinguished by their fat content, whether they have been heat-treated, whipped, etc. In many jurisdictions there are regulations for each type.

United States

In the United States, cream is usually sold as:
- Half and half (10.5–18% fat)
- Light, coffee, or table cream (18–30% fat)
- Medium cream (25% fat)
- Whipping or light Whipping cream (30–36% fat)

- Heavy Whipping cream (36% or more)
- Extra-heavy, double, or manufacturer's cream (38–40% or more), generally not available at retail except at some warehouse and specialty stores.

Not all grades are defined by all jurisdictions, and the exact fat content ranges vary. The above figures are based on the Code of Federal Regulations, Title 21, Part 131 and a small sample of state regulations.

Australia

In Australia, levels of fat in cream are not regulated, therefore labels are only under the control of the manufacturers. A general guideline is as follows:

Extra light (or 'lite'): 12–12.5% fat.

Light (or 'lite'): 18–20% fat.

Pure cream: 35–56% fat, without artificial thickeners.

Thickened cream: 35–36.5% fat, with added gelatine and/or other thickeners to give the cream a creamier texture, also possibly with stabilisers to aid the consistency of whipped cream (this would be the cream to use for whipped cream, not necessarily for cooking)

Single cream: Recipes calling for 'single cream' are referring to pure or thickened cream with about 35% fat.

Double cream: 48–60% fat.

Canada

Canadian cream definitions are similar to those used in the United States, except for that of "light cream." In Canada, "light cream" is low-fat cream, with 5% or 6% fat. Another form of cream available in Canada is "cereal cream", which is approximately midway between 5% cream and coffee cream in fat content.

Japan

In Japan, cream sold in supermarkets is usually between 46% and 48% butterfat.

Processing and Additives

Cream may have thickening agents and stabilisers added. Thickeners include sodium alginate, carrageenan, gelatine, sodium bicarbonate, tetrasodium pyrophosphate, and alginic acid). Other processing may be carried out. For example, cream has a tendency to produce oily globules (called "feathering") when added to coffee. The stability of the cream may be increased by increasing the non-fat solids content, which can be done by partial demineralisation and addition of sodium caseinate, although this is expensive.

Other Cream Products

Butter is made by churning cream to separate the butterfat and buttermilk. This can be done by hand or by machine.

Whipped cream is made by whisking or mixing air into cream with more than 30% fat, to turn the liquid cream into a soft solid. Nitrous oxide may also be used to make whipped cream.

Sour cream, common in many countries including the U.S. and Australia, is cream (12 to 16% or more milk fat) that has been subjected to a bacterial culture that produces lactic acid (0.5%+), which sours and thickens it. Crème fraîche (28% milk fat) is slightly soured with bacterial culture, but not as sour or as thick as sour cream. Mexican crema (or cream espesa) is similar to crème fraîche.

Smetana is a heavy cream product (35-40% milk fat) Central and Eastern European sour cream. Rjome or rømme is Norwegian sour cream containing 35% milk fat, similar to Icelandic rjómi.

Clotted cream, common in the United Kingdom, is cream that has been slowly heated to dry and thicken it, producing a very high-fat (55%) product. This is similar to Indian malai.

Cream as an Ingredient

Cream is used as an ingredient in many foods, including ice cream, many sauces, soups, stews, puddings, and some custard bases, and is also used for cakes. Irish cream is an alcoholic liqueur which blends cream with whiskey and coffee. Cream is also used in curries such as masala dishes. Cream (usually light/single cream or half and half) is often added to coffee. For cooking purposes, both single and double cream can be used in cooking, although the former can separate when heated, usually if there is a high acid content. Most UK chefs always use double cream or full-fat crème fraîche when cream is added to a hot sauce, to prevent any problem with it separating or "splitting". In sweet and savoury custards such as those found in flan fillings, crème brûlées and crème caramels, both types of cream are called for in different recipes depending on how rich a result is called for. It is useful to note that double cream can also be thinned down with water to make an approximation of single cream if necessary.

Other Items Called "Cream"

Many non-edible substances are called creams due merely to their consistency: shoe cream is runny, unlike waxy shoe polish; face cream is a cosmetic. There is generally no restriction on describing non-edible products as creams. Regulations in many jurisdictions restrict the use of the word *cream* for foods. Words such as *creme, kreme, creame,* or *whipped topping* are often used for products which cannot legally be called cream.

In some cases foods can be described as cream although they do not contain predominantly milk fats; for example in Britain "ice cream" does not have to be a dairy product (although it must be labelled "contains non-milk fat"), and salad cream is the customary name for a condiment that has been produced since the 1920s and need contain no cream.

Butter

Butter is a dairy product made by churning fresh or fermented cream or milk. It is generally used as a spread and a condiment, as well as in cooking applications, such as baking, sauce making, and pan frying. Butter consists of butterfat, water and milk proteins.

Most frequently made from cows' milk, butter can also be manufactured from the milk of other mammals, including sheep, goats, buffalo, and yaks. Salt, flavourings and preservatives are sometimes added to butter. Rendering butter produces clarified butter or *ghee*, which is almost entirely butterfat. Butter is a water-in-oil emulsion resulting from an inversion of the cream, an oil-in-water emulsion; the milk proteins are the emulsifiers.

Butter remains a solid when refrigerated, but softens to a spreadable consistency at room temperature, and melts to a thin liquid consistency at 32–35 °C (90–95 °F). The density of butter is 911 g/L (56.9 lb/ft^3).

It generally has a pale yellow colour, but varies from deep yellow to nearly white. Its unmodified colour is dependent on the animals' feed and is commonly manipulated with food colorings in the commercial manufacturing process, most commonly annatto or carotene.

Etymology

The word *butter* derives (via Germanic languages) from the Latin *butyrum*, which is the latinisation of the Greek *bouturon*. This may have been a construction meaning "cow-cheese", from *bous*, "ox, cow" + *turos*, "cheese", but perhaps this is a false etymology of a Scythian word. Nevertheless, the earliest attested form of the second stem, *turos* ("cheese"), is the Mycenaean Greek *tu-ro*, written in Linear B syllabic script.

The root word persists in the name butyric acid, a compound found in rancid butter and dairy products such as Parmesan cheese. Another possibility may be an extended derivation from Sanskrit *bhutari*, meaning "the enemy of evil spirits".

In general use, the term "butter" refers to the spread dairy product when unqualified by other descriptors. The word commonly is used to describe puréed vegetable or nut products such as peanut butter and almond butter. It is often applied to spread fruit products such as apple butter.

Fats such as cocoa butter and shea butter that remain solid at room temperature are also known as "butters". In addition to the act of applying butter being called "to butter", non-dairy items that have a dairy butter consistency may use "butter' to call that consistency to mind, including food items such as maple butter and Witch's butter and non-food items such as baby bottom butter, hyena butter, and rock butter.

Production

Unhomogenized milk and cream contain butterfat in microscopic globules. These globules are surrounded by membranes made of phospholipids (fatty acid emulsifiers) and proteins, which prevent the fat in milk from pooling together into a single mass. Butter is produced by agitating cream, which damages these membranes and allows the milk fats to conjoin, separating from the other parts of the cream.

Variations in the production method will create butters with different consistencies, mostly due to the butterfat composition in the finished product. Butter contains fat in three separate forms: free butterfat, butterfat crystals, and undamaged fat globules. In the finished product, different proportions of these forms result in different consistencies within the butter; butters with many crystals are harder than butters dominated by free fats.

Churning produces small butter grains floating in the water-based portion of the cream. This watery liquid is called buttermilk—although the buttermilk most common today is instead a directly fermented skimmed milk.

The buttermilk is drained off; sometimes more buttermilk is removed by rinsing the grains with water. Then the grains are "worked": pressed and kneaded together. When prepared manually, this is done using wooden boards called scotch hands. This consolidates the butter into a solid mass and breaks up embedded pockets of buttermilk or water into tiny droplets.

Commercial butter is about 80% butterfat and 15% water; traditionally made butter may have as little as 65% fat and 30% water. Butterfat consists of many moderate-sized, saturated hydrocarbon chain fatty acids. It is a triglyceride, an ester derived from glycerol and three fatty acid groups. Butter becomes rancid when these chains break down into smaller components, like butyric acid and diacetly. The density of butter is 0.911 g/cm^3 (0.527 oz/in^3), about the same as ice.

Types

Before modern factory butter making, cream was usually collected from several milkings and was therefore several days old and somewhat fermented by the time it was made into butter. Butter made from a fermented cream is known as cultured butter. During fermentation, the cream naturally sours as bacteria convert milk sugars into lactic acid.

The fermentation process produces additional aroma compounds, including diacetly, which makes for a fuller-flavoured and more "buttery" tasting product. Today, cultured butter is usually made from pasteurized cream whose fermentation is produced by the introduction of *Lactococcus* and *Leuconostoc* bacteria. Another method for producing cultured butter, developed in the early 1970s, is to produce butter from fresh cream and then incorporate bacterial cultures and lactic acid. Using this method, the cultured butter flavour grows as the butter is aged in cold storage.

For manufacturers, this method is more efficient, since aging the cream used to make butter takes significantly more space than simply storing the finished butter product. A method to make an artificial simulation of cultured butter is to add lactic acid and flavour compounds directly to the fresh-cream butter; while this more efficient process is claimed to simulate the taste of cultured butter, the product produced is not cultured but is instead flavoured.

Dairy products are often pasteurized during production to kill pathogenic bacteria and other microbes. Butter made from pasteurized fresh cream is called sweet cream butter. Production of sweet cream butter first became common in the 19th century, with the development of refrigeration and the mechanical cream separator. Butter made from fresh or cultured unpasteurized cream is called raw cream butter. Raw cream butter has a "cleaner" cream flavour, without the cooked-milk notes that pasteurization introduces.

Throughout Continental Europe, cultured butter is preferred, while sweet cream butter dominates in the United States and the United Kingdom. Therefore, cultured butter is sometimes labelled *European-style butter* in the United States. Commercial raw cream butter is virtually unheard-of in the United States.

Raw cream butter is generally only found made at home by consumers who have purchased raw whole milk directly from dairy farmers, skimmed the cream themselves, and made butter with it. It is rare in Europe as well. Several spreadable butters have been developed; these remain softer at colder temperatures and are therefore easier to use directly out of refrigeration. Some modify the make-up of the butter's fat through chemical manipulation of the finished product, some through manipulation of the cattle's feed, and some by incorporating vegetable oils into the butter. Whipped butter, another product designed to be more spreadable, is aerated via the incorporation of nitrogen gas—normal air is not used, because doing so would encourage oxidation and rancidity.

All categories of butter are sold in both salted and unsalted forms. Either granular salt or a strong brine are added to salted butter during processing. In addition to enhanced flavour, the addition of salt acts as a preservative.

The amount of butterfat in the finished product is a vital aspect of production. In the United States, products sold as "butter" are required to contain a minimum of 80% butterfat; in practice most American butters contain only slightly more than that, averaging around 81% butterfat. European butters generally have a higher ratio, which may extend up to 85%.

Clarified butter is butter with almost all of its water and milk solids removed, leaving almost-pure butterfat. Clarified butter is made by heating butter to its melting point and then allowing it to cool off; after settling, the remaining components separate by density. At the top, whey proteins form a skin which is removed, and the resulting butterfat is then poured off from the mixture of water and casein proteins that settle to the bottom.

Ghee is clarified butter which is brought to higher temperatures of around 120 °C (250 °F) once the water has cooked off, allowing the milk solids to brown. This process flavours the ghee, and also produces antioxidants which help protect it longer from rancidity. Because of this, ghee can keep for six to eight months under normal conditions.

Cream may be skimmed from whey instead of milk, as a by-product of cheese-making. Whey butter may be made from whey cream. Whey cream and butter have a lower fat content and taste more salty, tangy and "cheesy". They are also cheaper than "sweet" cream and butter.

European Butters

There are several butters produced in Europe with Protected geographical indications, these include:

- Beurre d'Ardenne, from Belgium
- Beurre d'Isigny, from France
- Beurre Charentes-Poitou (Which also includes: Beurre des Charentes and Beurre des Deux-Sèvres under the same classification), from France
- Beurre Rose, from Luxembourg
- Mantequilla de Soria, from Spain
- Mantega de l'Alt Urgell i la Cerdanya, from Spain.

History

The earliest butter would have been from sheep or goat's milk; cattle are not thought to have been domesticated for another thousand years. An ancient method of butter making, still used today in parts of Africa and the Near East, involves a goat skin half filled with milk, and inflated with air before being sealed. The skin is then hung with ropes on a tripod of sticks, and rocked until the movement leads to the formation of butter. In the Mediterranean climate, unclarified butter spoils quickly— unlike cheese it is not a practical method of preserving the nutrients of milk.

The ancient Greeks and Romans seemed to have considered butter a food fit more for the northern barbarians. A play by the Greek comic poet Anaxandrides refers to Thracians as *boutyrophagoi*; "butter-eaters". In *Natural History*, Pliny the Elder calls butter "the most delicate of food among barbarous nations", and goes on to describe its medicinal properties. Later, the physician Galen also described butter as a medicinal agent only.

Historian and linguist Andrew Dalby says that most references to butter in ancient Near Eastern texts should more correctly be translated as ghee. Ghee is mentioned in the Periplus of the Erythraean Sea as a typical trade article around the 1st century CE Arabian Sea, and Roman geographer Strabo describes it as a commodity of Arabia and Sudan. In India, ghee has been a symbol of purity and an offering to the gods—especially Agni, the Hindu god of fire—for more than 3000 years; references to ghee's sacred nature appear numerous times in the Rig Veda, circa 1500–1200 BCE. The tale of the child Krishna stealing butter remains a popular children's story in India today. Since India's prehistory, ghee has been both a staple food and used for ceremonial purposes such as fueling holy lamps and funeral pyres.

Middle Ages

The cooler climates of northern Europe allowed butter to be stored for a longer period before it spoiled. Scandinavia has the oldest tradition in Europe of butter export trade, dating at least to the 12th century. After the fall of Rome and through much of the Middle Ages, butter was a common food across most of Europe, but one with a low reputation, and was consumed principally by peasants.

Butter slowly became more accepted by the upper class, notably when the early 16th century Roman Catholic Church allowed its consumption during Lent. Bread and butter became common fare among the middle class and the English, in particular, gained a reputation for their liberal use of melted butter as a sauce with meat and vegetables.

In antiquity, butter was used for fuel in lamps as a substitute for oil. The *Butter Tower* of Rouen Cathedral was erected in the early 16th century when Archbishop Georges d'Amboise authorized the burning of butter instead of oil, which was scarce at the time, during Lent. Across northern Europe, butter was sometimes treated in a manner unheard-of today: it was packed into barrels (firkins) and buried in peat bogs, perhaps for years. Such "bog butter" would develop a strong flavour as it aged, but remain edible, in large part because of the unique cool, airless, antiseptic and acidic environment of a peat bog. Firkins of such buried butter are a common archaeological find in Ireland; the Irish National Museum has some containing "a grayish cheese-like substance, partially hardened, not much like butter, and quite free from putrefaction." The practice was most common in Ireland in the 11th–14th centuries; it ended entirely before the 19th century.

Industrialization

Like Ireland, France became well-known for its butter, particularly in Normandy and Brittany. By the 1860s, butter had become so in demand in France that Emperor Napoleon III offered prize money for an inexpensive substitute to supplement France's inadequate butter supplies.

A French chemist claimed the prize with the invention of margarine in 1869. The first margarine was beef tallow flavoured with milk and worked like butter; vegetable margarine followed after the development of hydrogenated oils around 1900. Until the 19th century, the vast majority of butter was made by hand, on farms. The first butter factories appeared in the United States in the early 1860s, after the successful introduction of cheese factories a decade earlier. In the late 1870s, the centrifugal cream separator was introduced, marketed most successfully by Swedish engineer Carl Gustaf Patrik de Laval. This dramatically sped up the butter-making process by eliminating the slow step of letting cream naturally rise to the top of milk. Initially, whole milk was shipped to the butter factories, and the cream separation took place there.

Soon, though, cream-separation technology became small and inexpensive enough to introduce an additional efficiency: the separation was accomplished on the farm, and the cream alone shipped to the factory. By 1900, more than half the butter produced in the United States was factory made; Europe followed suit shortly after. In 1920, Otto Hunziker authored *The Butter Industry, Prepared for Factory, School and Laboratory*, a well-known text in the industry that enjoyed at least three editions (1920, 1927, 1940).

As part of the efforts of the American Dairy Science Association, Professor Hunziker and others published articles regarding: causes of tallowiness (an odour defect, distinct from rancidity, a taste defect); mottles (an aesthetic issue related to uneven colour); introduced salts; the impact of creamery metals and liquids; and acidity measurement. These and other ADSA publications helped standardize practices internationally. Per capita butter consumption declined in most western nations during the 20th century, in large part because of the rising popularity of margarine, which is less expensive and, until recent years, was perceived as being healthier. In the United States, margarine consumption overtook butter during the 1950s and it is still the case today that more margarine than butter is eaten in the U.S. and the EU.

Shape of Butter Sticks

In the United States, butter is usually produced in 4-ounce sticks, wrapped in waxed or foiled paper and sold four to a one-pound carton. This practice is believed to have originated in 1907, when Swift and Company began packaging butter in this manner for mass distribution.

Due to historical differences in butter printers (the machines which cut and package butter), these sticks are commonly produced in two different shapes:

- The dominant shape east of the Rocky Mountains is the Elgin, or Eastern-pack shape, named for a dairy in Elgin, Illinois. The sticks are $4\frac{3}{4}$ inches long and $1\frac{1}{4}$ inches (121 mm × 32 mm) wide and are typically sold stacked two by two in elongated cube-shaped boxes.

- West of the Rocky Mountains, butter printers standardized on a different shape that is now referred to as the Western-pack shape. These butter sticks are $3\frac{1}{8}$ inches long and $1\frac{1}{2}$ inches wide (80 mm × 38 mm) and are usually sold with four sticks packed side-by-side in a flat, rectangular box.

Both sticks contain the same amount of butter, although most butter dishes are designed for Elgin-style butter sticks.

The stick's wrapper is usually marked off as eight tablespoons (120 ml/4.2 imp fl oz; 4.1 US fl oz); the actual volume of one stick is approximately nine tablespoons (130 ml/4.6 imp fl oz; 4.4 US fl oz).

World Wide

In 1997, India produced 1,470,000 metric tons (1,620,000 short tons) of butter, most of which was consumed domestically. Second in production was the United States (522,000 t/575,000 short tons), followed by France (466,000 t/514,000 short tons), Germany (442,000 t/ 487,000 short tons), and New Zealand (307,000 t/338,000 short tons). France ranks first in per capita butter consumption with 8 kg per capita per year. In terms of absolute consumption, Germany was second after India, using 578,000 metric tons (637,000 short tons) of butter in 1997, followed by France (528,000 t/582,000 short tons), Russia (514,000 t/567,000 short tons), and the United States (505,000 t/ 557,000 short tons). New Zealand, Australia, and the Ukraine are among the few nations that export a significant percentage of the butter they produce.

Different varieties are found around the world. *Smen* is a spiced Moroccan clarified butter, buried in the ground and aged for months or years. Yak butter is important in Tibet; *tsampa*, barley flour mixed with yak butter, is a staple food. Butter tea is consumed in the Himalayan regions of Tibet, Bhutan, Nepal and India. It consists of tea served with intensely flavoured—or "rancid"—yak butter and salt. In African and Asian developing nations, butter is traditionally made from sour milk rather than cream. It can take several hours of churning to produce workable butter grains from fermented milk.

Storage and Cooking

Normal butter softens to a spreadable consistency around 15 °C (60 °F), well above refrigerator temperatures. The "butter compartment" found in many refrigerators may be one of the warmer sections inside, but it still leaves butter quite hard.

Until recently, many refrigerators sold in New Zealand featured a "butter conditioner", a compartment kept warmer than the rest of the refrigerator—but still cooler than room temperature—with a small heater. Keeping butter tightly wrapped delays rancidity, which is hastened by exposure to light or air, and also helps prevent it from picking up other odors. Wrapped butter has a shelf life of several months at refrigerator temperatures.

"French butter dishes" or "Acadian butter dishes" involve a lid with a long interior lip, which sits in a container holding a small amount of water. Usually the dish holds just enough water to submerge the interior lip when the dish is closed. Butter is packed into the lid. The water acts as a seal to keep the butter fresh, and also keeps the butter from overheating in hot temperatures. This allows butter to be safely stored on the countertop for several days without spoilage.

Once butter is softened, spices, herbs, or other flavoring agents can be mixed into it, producing what is called a *compound butter* or *composite butter* (sometimes also called *composed butter*). Compound butters can be used as spreads, or cooled, sliced, and placed onto hot food to melt into a sauce. Sweetened compound butters can be served with desserts; such hard sauces are often flavoured with spirits.

Melted butter plays an important role in the preparation of sauces, most obviously in French cuisine. *Beurre noisette* (hazelnut butter) and *Beurre noir* (black butter) are sauces of melted butter cooked until the milk solids and sugars have turned golden or dark brown; they are often finished with an addition of vinegar or lemon juice. Hollandaise and béarnaise sauces are emulsions of egg yolk and melted butter; they are in essence mayonnaises made with butter instead of oil. Hollandaise and béarnaise sauces are stabilized with the powerful emulsifiers in the egg yolks, but butter itself contains enough emulsifiers—mostly remnants of the fat globule membranes—to form a stable emulsion on its own.

Beurre blanc (white butter) is made by whisking butter into reduced vinegar or wine, forming an emulsion with the texture of thick cream. *Beurre monté* (prepared butter) is melted but still emulsified butter; it lends its name to the practice of "mounting" a sauce with butter: whisking cold butter into any water-based sauce at the end of cooking, giving the sauce a thicker body and a glossy shine—as well as a buttery taste. In Poland, the butter lamb (*Baranek wielkanocny*) is a traditional addition to the Easter Meal for many Polish Catholics. Butter is shaped into a lamb either by hand or in a lamb-shaped mould. Butter is also used to make edible decorations to garnish other dishes.

Butter is used for sautéing and frying, although its milk solids brown and burn above 150 °C (250 °F)—a rather low temperature for most applications. The smoke point of butterfat is around 200 °C (400 °F), so clarified butter or ghee is better suited to frying. Ghee has always been a common frying medium in India, where many avoid other animal fats for cultural or religious reasons.

Butter fills several roles in baking, where it is used in a similar manner as other solid fats like lard, suet, or shortening, but has a flavour that may better complement sweet baked goods. Many cookie doughs and some cake batters are leavened, at least in part, by creaming butter and sugar together, which introduces air bubbles into the butter. The tiny bubbles locked within the butter expand in the heat of baking and aerate the cookie or cake. Some cookies like shortbread may have no other source of moisture but the water in the butter. Pastries like pie dough incorporate pieces of solid fat into the dough, which become flat layers of fat when the dough is rolled out.

During baking, the fat melts away, leaving a flaky texture. Butter, because of its flavour, is a common choice for the fat in such a dough, but it can be more difficult to work with than shortening because of its low melting point. Pastry makers often chill all their ingredients and utensils while working with a butter dough. Butter also has many non-culinary, traditional uses which are specific to certain cultures. For instance, in North America, applying butter to the handle of a door is a common prank on April Fools' Day.

Health and Nutrition

Butter, Unsalted

Nutritional Value Per 100 g (3.5 oz)

Energy	2,999 kJ (717 kcal)
Carbohydrates	0 g
Fat	81 g
Saturated	51 g
Monounsaturated	21 g
Polyunsaturated	3 g
Protein	1 g
Vitamin A equiv.	684 ìg (76%)
Vitamin D	60 IU (15%)
Vitamin E	2.32 mg (15%)
Cholesterol	215 mg

Fat percentage can vary.

Percentages are relative to US recommendations for adults.

Source: USDA Nutrient database

According to USDA figures, one tablespoon of butter (14 grams / 0.5 ounces) contains 420 kilojoules (100 kcal), all from fat, 11 grams (0.4 oz) of fat, of which 7 grams (0.25 oz) are saturated fat, and 30 milligrams (0.46 gr) of cholesterol. Butter consists mostly of saturated fat and is a significant source of cholesterol.

For these reasons butter is considered by some to be a contributor to health problems, especially heart disease. Margarine was recommended as a substitute, since it is higher in unsaturated fat and contains little or no cholesterol, but in recent years, it has been shown that the trans fats contained in partially hydrogenated oils used in typical margarines significantly raise undesirable LDL cholesterol levels as well.

Trans-fat–free margarines have since been developed. Proponents of the consumption of organic butter, such as the nutritionist Mary Enig, state that since butter is nutritious and "is rich in short and medium chain fatty acids," this can have a positive effect on health and prevent disease.

Butter contains only traces of lactose, so moderate consumption of butter is not a problem for the lactose intolerant. People with milk allergies need to avoid butter, which contains enough of the allergy-causing proteins to cause reactions. Butter can perform a useful role in dieting by providing satiety. A small amount added to low fat foods such as vegetables may stave off feelings of hunger.

Icecream and Related Frozen Dairy Desserts

Ice cream is a frozen dessert usually made from dairy products, such as milk and cream, and often combined with fruits or other ingredients and flavours. Most varieties contain sugar, although some are made with other sweeteners. In some cases, artificial flavourings and colorings are used in addition to (or in replacement of) the natural ingredients. This mixture is stirred slowly while cooling to prevent large ice crystals from forming; the result is a smoothly textured ice cream.

The meaning of the term ice cream varies from one country to another. Terms like frozen custard, frozen yogurt, sorbet, gelato and others are used to distinguish different varieties and styles. In some countries, like the USA, the term ice cream applies only to a specific variety, and their governments regulate the commercial use of all these terms based on quantities of ingredients. In others, like Italy and Argentina, one word is used for all the variants. Alternatives made from soy milk, rice milk, and goat milk are available for those who are lactose intolerant or have an allergy to dairy protein, or in the case of soy and rice milk, for those who want to avoid animal products.

History

Precursors of Ice Cream

In the Persian Empire, people would pour grape juice concentrate over snow - in a bowl - and eat this as a treat. In particular this was consumed when the weather was hot. Either snow would be saved in the cool-keeping underground chambers known as "yakhchal" or taken from fresh snow that may still have remained at the top of the mountains by the summer capital - Hagmatana, Ecbatana or Hamedan of today. In 400 BC, the Persians went further and invented a special chilled food, made of rose water and vermicelli which was served to royalty during summers. The ice was mixed with saffron, fruits, and various other flavours. Ancient civilizations have served ice for cold foods for thousands of years. The BBC reports that a frozen mixture of milk and rice was used in China around 200 BC. The Roman Emperor Nero (37–68) had ice brought from the mountains and combined with fruit toppings. These were some early chilled delicacies.

Arabs were the first to use milk as a major ingredient in its production, sweeten the ice cream with sugar rather than fruit juices, as well as perfect ways for its commercial production. As early as the 10th century, ice cream was widespread amongst many of the Arab world's major cities, such as Baghdad, Damascus and Cairo. Their version of ice cream was produced from milk or cream and often some yoghurt similar to Ancient Greek recipes, flavoured with rosewater as well as dried fruits and nuts. It is believed that this was based on older Ancient Arab, Mesopotamian, Greek or Roman recipes, which were probably the first and precursors to Persian faloodeh.

In 62 AD, the Roman emperor Nero sent slaves to the Apennine mountains to collect snow to be flavoured with honey and nuts.

Maguelonne Toussaint-Samat asserts in her *History of Food*, "the Chinese may be credited with inventing a device to make sorbets and ice cream. They poured a mixture of snow and saltpetre over the exteriors of containers filled with syrup, for, in the same way as salt raises the boiling-point of water, it lowers the freezing-point to below zero." (Toussaint does not provide historical documentation for this.) Some distorted accounts claim that in the age of Emperor Yingzong, Song Dynasty (960-1279) of China, was written by the poet Yang Wanli. Actually, and has nothing to do with ice cream. It has also been claimed that, in the Yuan Dynasty, Kublai Khan enjoyed ice cream and kept it a royal secret until Marco Polo visited China and took the technique of making ice cream to Italy. Others have argued that the Chinese didn't drink milk during that period, whereas the Italians had arguably been making something resembling ice cream before Marco Polo returned to Italy. In any case, no known ice cream recipes appear to stem from ancient Chinese sources.

In the sixteenth century, the Mughal emperors used relays of horsemen to bring ice from the Hindu Kush to Delhi, where it was used in fruit sorbets.

When Italian duchess Catherine de' Medici married the duc d'Orléans in 1533, she is said to have brought with her Italian chefs who had recipes for flavoured ices or sorbets, and introduced them in France.

One hundred years later, Charles I of England was supposedly so impressed by the "frozen snow", he offered his own ice cream maker a lifetime pension in return for keeping the formula secret, so ice cream could be a royal prerogative. There is no historical evidence to support these legends, which first appeared during the 19th century.

The first recipe for flavoured ices in French appears in 1674, in Nicholas Lemery's *Recueil de curiositéz rares et nouvelles de plus admirables effets de la nature*. Recipes for *sorbetti* saw publication in the 1694 edition of Antonio Latini's *Lo Scalco alla Moderna* (The Modern Steward). Recipes for flavoured ices begin to appear in François Massialot's *Nouvelle Instruction pour les Confitures, les Liqueurs, et les Fruits* starting with the 1692 edition. Massialot's recipes result in a coarse, pebbly texture. Latini claims that the results of his recipes should have the fine consistency of sugar and snow.

True Ice Cream

Ice cream recipes first appear in 18th century England and America. A recipe for ice cream was published in *Mrs. Mary Eales's Receipts* in London 1718.

To ice CREAM. Take Tin Ice-Pots, fill them with any Sort of Cream you like, either plain or sweeten'd, or Fruit in it; shut your Pots very close; to six Pots you must allow eighteen or twenty Pound of Ice, breaking the Ice very small; there will be some great Pieces, which lay at the Bottom and Top: You must have a Pail, and lay some Straw at the Bottom; then lay in your Ice, and put in amongst it a Pound of Bay-Salt; set in your Pots of Cream, and lay Ice and Salt between every Pot, that they may not touch; but the Ice must lie round them on every Side; lay a good deal of Ice on the Top, cover the Pail with Straw, set it in a Cellar where no Sun or Light comes, it will be froze in four Hours, but it may stand longer; than take it out just as you use it; hold it in your Hand and it will slip out.

When you would freeze any Sort of Fruit, either Cherries, Raspberries, Currants, or Strawberries, fill your Tin-Pots with the Fruit, but as hollow as you can; put to them Lemmonade, made with Spring-Water and Lemmon-Juice sweeten'd; put enough in the Pots to make the Fruit hang together, and put them in Ice as you do Cream.

The earliest reference to ice cream given by the *Oxford English Dictionary* is from 1744, reprinted in a magazine in 1877. *1744 in Pennsylvania Mag. Hist. & Biogr. (1877) I. 126 Among the rarities..was some fine ice cream, which, with the strawberries and milk, eat most deliciously.*

The 1751 edition of *The Art of Cookery made Plain and Easy* by Hannah Glasse features a recipe for ice cream. OED gives her recipe: *H. GLASSE Art of Cookery (ed. 4) 333 (heading) To make Ice Cream..set it [sc. the cream] into the larger Bason. Fill it with Ice, and a Handful of Salt.* 1768 saw the publication of *L'Art de Bien Faire les Glaces d'Office* by M. Emy, a cookbook devoted entirely to recipes for flavoured ices and ice cream.

Ice cream was introduced to the United States by Quaker colonists who brought their ice cream recipes with them. Confectioners sold ice cream at their shops in New York and other cities during the colonial era. Ben Franklin, George Washington, and Thomas Jefferson were known to have regularly eaten and served ice cream. First Lady Dolley Madison is also closely associated with the early history of ice cream in the United States. One respected history of ice cream states that, as the wife of U.S. President James Madison, she served ice cream at her husband's Inaugural Ball in 1813.

Around 1832, Augustus Jackson, an African American confectioner, not only created multiple ice cream recipes, but he also invented a superior technique to manufacture ice cream.

In 1843, Nancy Johnson of Philadelphia was issued the first U.S. patent for a small-scale handcranked ice cream freezer. The invention of the ice cream soda gave Americans a new treat, adding to ice cream's popularity. This cold treat was probably invented by Robert Green in 1874, although there is no conclusive evidence to prove his claim.

The ice cream sundae originated in the late 19th century. Several men claimed to have created the first sundae, but there is no conclusive evidence to back up any of their stories. Some sources say that the sundae was invented to circumvent blue laws, which forbade serving sodas on Sunday. Towns claiming to be the birthplace of the sundae include Buffalo, New York; Two Rivers, Wisconsin; Ithaca, New York; and Evanston, Illinois. Both the ice cream cone and banana split became popular in the early 20th century. Several food vendors claimed to have invented the ice cream cone at the 1904 World's Fair in St. Louis, MO. Europeans were eating cones long before 1904.

In the UK, ice cream remained an expensive and rare treat, until large quantities of ice began to be imported from Norway and the US in the mid Victorian era. A Swiss-Italian businessman, Carlo Gatti, opened the first ice cream stall outside Charing Cross station in 1851, selling scoops of ice cream in shells for one penny.

The history of ice cream in the 20th century is one of great change and increases in availability and popularity. In the United States in the early 20th century, the ice cream soda was a popular treat at the soda shop, the soda fountain, and the ice cream parlor. During American

Prohibition, the soda fountain to some extent replaced the outlawed alcohol establishments such as bars and saloons.

Ice cream became popular throughout the world in the second half of the 20th century after cheap refrigeration became common. There was an explosion of ice cream stores and of flavours and types. Vendors often competed on the basis of variety. Howard Johnson's restaurants advertised "a world of 28 flavours." Baskin-Robbins made its 31 flavours ("one for every day of the month") the cornerstone of its marketing strategy. The company now boasts that it has developed over 1000 varieties.

One important development in the 20th century was the introduction of soft ice cream. A chemical research team in Britain (of which a young Margaret Thatcher was a member) discovered a method of doubling the amount of air in ice cream, which allowed manufacturers to use less of the actual ingredients, thereby reducing costs. It made possible the soft ice cream machine in which a cone is filled beneath a spigot on order. In the United States, Dairy Queen, Carvel, and Tastee-Freez pioneered in establishing chains of soft-serve ice cream outlets.

Technological innovations such as these have introduced various food additives into ice cream, notably the stabilizing agent gluten, to which some people have an intolerance. Recent awareness of this issue has prompted a number of manufacturers to start producing gluten-free ice cream. The 1980s saw a return of the older, thicker ice creams being sold as "premium" and "superpremium" varieties under brands such as Ben & Jerry's and Häagen-Dazs.

Production

Before the development of modern refrigeration, ice cream was a luxury reserved for special occasions. Making it was quite laborious; ice was cut from lakes and ponds during the winter and stored in holes in the ground, or in wood-frame or brick ice houses, insulated by straw. Many farmers and plantation owners, including U.S. Presidents George Washington and Thomas Jefferson, cut and stored ice in the winter for use in the summer. Frederic Tudor of Boston turned ice harvesting and shipping into a big business, cutting ice in New England and shipping it around the world. Ice cream was made by hand in a large bowl placed inside a tub filled with ice and salt.

This was called the pot-freezer method. French confectioners refined the pot-freezer method, making ice cream in a sorbetière (a covered pail with a handle attached to the lid). In the pot-freezer method, the temperature of the ingredients is reduced by the mixture of crushed ice and salt. The salt water is cooled by the ice, and the action of the salt on the ice causes it to (partially) melt, absorbing latent heat and bringing the mixture below the freezing point of pure water. The immersed container can also make better thermal contact with the salty water and ice mixture than it could with ice alone. The hand-cranked churn, which also uses ice and salt for cooling, replaced the pot-freezer method.

The exact origin of the hand-cranked freezer is unknown, but the first U.S. patent for one was #3254 issued to Nancy Johnson on September 9, 1843. The hand-cranked churn produced smoother ice cream than the pot freezer and did it quicker. Many inventors patented improvements on Johnson's design. In Europe and early America, ice cream was made and sold by small businesses, mostly confectioners and caterers. Jacob Fussell of Baltimore, Maryland was the first to manufacture ice cream on a large scale.

Fussell bought fresh dairy products from farmers in York County, Pennsylvania, and sold them in Baltimore. An unstable demand for his dairy products often left him with a surplus of cream, which he made into ice cream. He built his first ice-cream factory in Seven Valleys, Pennsylvania, in 1851. Two years later, he moved his factory to Baltimore. Later, he opened factories in several other cities and taught the business to others, who operated their own plants. Mass production reduced the cost of ice cream and added to its popularity.

The development of industrial refrigeration by German engineer Carl von Linde during the 1870s eliminated the need to cut and store natural ice and when the continuous-process freezer was perfected in 1926, it allowed commercial mass production of ice cream and the birth of the modern ice cream industry.

The most common method for producing ice cream at home is to use an ice cream maker, in modern times generally an electrical device that churns the ice cream mixture while cooled inside a household freezer, or using a solution of pre-frozen salt and water, which gradually melts while the ice cream freezes. Some more expensive models have an inbuilt freezing element. A newer method of making home-made ice cream is to add liquid nitrogen to the mixture while stirring it using a spoon or spatula. Some ice cream recipes call for making a custard, folding in whipped cream, and immediately freezing the mixture.

Commercial Delivery

Ice cream can be mass-produced and thus is widely available in developed parts of the world. Ice cream can be purchased in large cartons (vats and squrounds) from supermarkets and grocery stores, in smaller quantities from ice cream shops, convenience stores, and milk bars, and in individual servings from small carts or vans at public events.

In Turkey and Australia, ice cream is sometimes sold to beach-goers from small powerboats equipped with chest freezers. Some ice cream distributors sell ice cream products from traveling refrigerated vans or carts (commonly referred to in the US as "ice cream trucks"), sometimes equipped with speakers playing children's music. Traditionally, ice cream vans in the United Kingdom make a music box noise rather than actual music.

Dietary

Ice cream may have the following composition:

- greater than 10% milkfat and usually between 10% and as high as 16% fat in some premium ice creams
- 9 to 12% milk solids-not-fat: this component, also known as the serum solids, contains the proteins (caseins and whey proteins) and carbohydrates (lactose) found in milk
- 12 to 16% sweeteners: usually a combination of sucrose and glucose-based corn syrup sweeteners
- 0.2 to 0.5% stabilisers and emulsifiers
- 55% to 64% water which comes from the milk or other ingredients.

These compositions are percentage by weight. Since ice cream can contain as much as half air by volume, these numbers may be reduced by as much as half if cited by volume. In terms of dietary considerations, The percentages by weight are more relevant. Even the low fat products have high caloric content: Ben and Jerry's No Fat Vanilla Fudge contains 150 calories per half cup due to its high sugar content.

Ice Cream around the World

Ice Cream Cone

Mrs Marshall's Cookery Book, published in 1888, endorsed serving ice cream in cones, but the idea definitely predated that. Agnes Marshall was a celebrated cookery writer of her day and helped to popularise ice cream. She patented and manufactured an ice cream maker and was the first person to suggest using liquefied gases to freeze ice cream after seeing a demonstration at the Royal Institution. Reliable evidence proves that ice cream cones were served in the 19th century, and their popularity increased greatly during the St. Louis World's Fair in 1904. According to legend, at the World's Fair an ice cream seller had run out of the cardboard dishes used to put ice cream scoops in, so they could not sell any more produce. Next door to the ice cream booth was a Syrian waffle booth, unsuccessful due to intense heat; the waffle maker offered to make cones by rolling up his waffles and the new product sold well, and was widely copied by other vendors.

Other Frozen Desserts

The following is a partial list of ice cream-like frozen desserts and snacks:

- Ais kacang: a dessert in Malaysia and Singapore made from shaved ice, syrup, and boiled red bean and topped with evaporated milk. Sometimes, other small ingredients like raspberries and durians are added in too.

- Dondurma: Turkish ice cream, made of salep and mastic resin

- Frozen custard: at least 10% milk fat and at least 1.4% egg yolk and much less air beaten into it, similar to Gelato, fairly rare. Known in Italy as Semifreddo.

- Frozen yogurt: a low fat or fat free alternative made with yogurt

- Gelato: an Italian frozen dessert having a lower milk fat content than ice cream and stabilised with ingredients such as eggs.

- Halo-halo: a popular Filipino dessert that is a mixture of shaved ice and milk to which are added various boiled sweet beans and fruits, and served cold in a tall glass or bowl.

- Ice milk: less than 10% milk fat and lower sweetening content, once marketed as "ice milk" but now sold as *low-fat ice cream* in the United States.

- Ice pop (or lolly): frozen fruit puree, fruit juice, or flavoured sugar water on a stick or in a flexible plastic sleeve.

- Kulfi: Believed to have been introduced to South Asia by the Mughal conquest in the 16th century; its origins trace back to the cold snacks and desserts of Arab and Mediterranean cultures.

- Mellorine: non-dairy, with vegetable fat substituted for milk fat
- Parevine: Kosher non-dairy frozen dessert established in 1969 in New York
- Sherbet: 1–2% milk fat and sweeter than ice cream.
- Sorbet: fruit puree with no dairy products
- Snow cones, made from balls of crushed ice topped with sweet syrup served in a paper cone, are consumed in many parts of the world. The most common places to find snow cones in the United States are at amusement parks.
- Maple toffee: A popular springtime treat in maple-growing areas is maple toffee, where maple syrup boiled to a concentrated state is poured over fresh snow congealing in a toffee-like mass, and then eaten from a wooden stick used to pick it up from the snow.

Using Liquid Nitrogen

Using liquid nitrogen to freeze ice cream is an old idea and has been used for many years to harden ice cream. The use of liquid nitrogen in the primary freezing of ice cream, that is to effect the transition from the liquid to the frozen state without the use of a conventional ice cream freezer, has only recently started to see commercialization.

Some commercial innovations have been documented in the National Cryogenic Society Magazine "Cold Facts". The most noted brands are Dippin' Dots, Blue Sky Creamery, Project Creamery, and Sub Zero Cryo Creamery.

The preparation results in a column of white condensed water vapor cloud, reminiscent of popular depictions of witches' cauldrons. The ice cream, dangerous to eat while still "steaming," is allowed to rest until the liquid nitrogen is completely vaporised. Sometimes ice cream is frozen to the sides of the container, and must be allowed to thaw.

Making ice cream with liquid nitrogen has advantages over conventional freezing. Due to the rapid freezing, the crystal grains are smaller, giving the ice cream a creamier texture, and allowing one to get the same texture by using less milkfat. Such ice crystals will grow very quickly via the processes of recrystallization thus obviating the original benefits unless steps are taken to inhibit ice crystal growth.

Gelato

Gelato is Italy's regional variant of ice cream. As such, gelato is made with some of the same ingredients as most other frozen dairy desserts. Milk, cream, various sugars, flavoring including fruit and nut purees are the main ingredients.

Gelato is different from some other ice creams because it has a lower butterfat content. Gelato typically contains 4-8% butterfat, versus 14% for many other ice creams. Gelato generally has slightly lower sugar content, averaging between 16-22% versus approximately 21% for most ice creams.

Non-fat milk is added as a solid. The sugar content in gelato is precisely balanced with the water content to act as an anti-freeze to prevent the gelato from freezing solid. Types of sugar used include sucrose, dextrose, and invert sugar to control apparent sweetness. Typically, gelato and Italian sorbet contain a stabilizing base. Egg yolks are used in yellow custard-based gelato flavours, including zabaione and creme caramel.

The mixture for gelato is typically made using a hot process, which includes pasteurization. White base is heated to 85°C (185°F). Heating the mix to 90°C (194°F) is essential for chocolate gelato, which is traditionally flavoured with cocoa powder. Yellow custard base, which contains egg yolks, is heated to 65°C (149°F). The gelato mix must age for several hours after pasteurization is complete for the milk proteins to hydrate, or bind, with water in the mix. This hydration reduces the size of the ice crystals, making a smoother texture in the final product. A non-traditional cold mix process is popular among some gelato makers in the United States.

Unlike most commercial ice creams in the United States, which are frozen with a continuous assembly line freezer, gelato is frozen very quickly in individual small batches in a batch freezer. The batch freezer incorporates air or overage into the mix as it freezes. Unlike most American-style ice creams, which can have an overage of up to 50%, gelato generally has between 20% and 35% overage. This results in a denser product with more intense flavour than many U.S. style ice creams. U.S. style ice cream, with a higher fat content, can be stored in a freezer for months. High-quality artisan gelato holds its peak flavour and texture (from delicate ice crystals) only for several days, even when stored carefully at the proper temperature. This is why *gelaterias* typically make their own gelato on the premises or nearby.

History

The history of gelato dates back to frozen desserts served in ancient Rome and Egypt made from ice and snow brought down from mountaintops and preserved below ground. Later, gelato appeared during banquets at the Medici court in Florence.

In fact, the Florentine cook Bernardo Buontalenti is said to have invented modern ice cream in 1565, as he presented his recipe and his innovative refrigerating techniques to Catherine de' Medici. She in turn brought the novelty to France, where in 1686 the Sicilian fisherman Francesco Procopio dei Coltelli perfected the first ice cream machine.

The popularity of gelato among larger shares of the population however only increased in the 1920s-1930s as in the northern Italian city of Varese, where the first mobile gelato cart was developed.

Overview

Gelato is typically flavoured with fresh fruit purees, cocoa and/ or nut pastes. If other ingredients such as chocolate flakes, nuts, small confections, cookies, or biscuits are added, they are added after the gelato is frozen. Gelato made with fresh fruit, sugar, and water and without dairy ingredients is sorbet. Use of the word sorbetto is a common affectation in the United States. In Italy, the word sorbetto refers to a cocktail.

Stracciatella

Stracciatella (from Italian *stracciato*, "torn apart") is an Italian egg-drop soup usually said to be "alla Romana" ("the way it's done in Rome"), but also popular in Marche and Emilia Romagna. It is prepared by beating eggs and adding grated parmesan cheese, salt, pepper, nutmeg, and sometimes semolina, and then adding this mixture to boiling broth. The broth is set whirling first with a whisk, and the beaten egg mixture added in a slow stream to produce the *stracciatelle* ("little shreds") of cooked egg in the broth, which is clarified by the process.

In Gelato

In Italy and Germany, gelato with a vanilla base and chocolate shavings is also called stracciatella or stracciatella ice. It is somewhat analogous to chocolate chip ice cream in North America though the chocolate is intended to be less chunky and more integrated with the gelato.

In Cheese

A particular kind of mozzarella (soft cheese) is also called stracciatella. Stracciatella is used as stuffing for the burrata from the Murgia region in Puglia. It is made with torn pieces of mozzarella and cream.

Frozen Custard

Frozen custard is a cold dessert similar to ice cream, made with eggs in addition to cream and sugar.

In the United States the Food and Drug Administration requires products marketed as frozen custard to contain at least 10 percent milkfat and 1.4 percent egg yolk solids. If it has less egg yolk solids, it is considered ice cream. In the United Kingdom frozen custard is not differentiated from other frozen desserts. Instead, if given a name other than frozen desserts they may be referred to as ice creams.

History

Most early ice cream recipes would be called frozen custard today, because of naming conventions. The modern term of frozen custard is primarily used as a method of brand differentiation and to comply with FDA requirements. One claim traces the invention of the term frozen custard to Coney Island in 1919. However, recipes for custard based ices existed before then, especially in France. In some areas of the United States, "frozen custard" or "custard" has become a synonym for soft serve.

Home Creation

The vast majority of ice cream desserts created at home are frozen custards. Despite commercialization of a limited number of flavours, home cooks can make frozen custard into innumerable flavours. Extra egg-yolk and a brief heating is used in order to make a custard base for the ice. This replaces commercially used ice cream additives such as gluten in order to create the same consistency. It is possible that the heating of the cream containing milk was initially done as a form of pasteurization, with the side effect of creating a custard. At home, frozen custard can be made with a standard ice cream maker or with the freeze and scrape method.

Commercial Creation

Using a process called overrun, air is blended into the mixture of ingredients until its volume increases by approximately 20%. By comparison, ice cream may have an overrun as large as 100%, meaning half of the final product is composed of air. The high percentage of butterfat and egg yolk gives frozen custard a thick, creamy texture and a smoother consistency than ice cream. Frozen custard can be served at –8°C (18°F), warmer than the –12°C (10°F) at which ice cream is served, in order to make a soft serve product.

Another difference between commercially produced frozen custard and commercial ice cream is the way the custard is frozen. The mix enters a refrigerated tube and, as it freezes, blades scrape the product cream off the barrel walls.

The now frozen custard is discharged directly into containers from which it can be served. The speed with which the product leaves the barrel minimizes the amount of air in the product but more importantly ensures that the ice crystals formed are very small. Frozen custard is usually prepared fresh at the place of sale, rather than stored; however, it is occasionally available in supermarkets or by mail order.

Generally, modern frozen custard stands provide only three different flavours per day: vanilla, chocolate, and a unique "flavour of the day." The older vintage custard stands tend to have a dozen or so standard flavours. Some locations try to emulate the Coney Island feel (sometimes cited as the birthplace of the term frozen custard) by serving other diner foods along with their premium dessert.

Semifreddo, a Class of Semi-frozen Desserts

Semifreddo (Italian: "half cold") is a class of semi-frozen desserts, typically ice-cream cakes, semi-frozen custards, and certain fruit tarts. It has the texture of frozen mousse because it is usually produced by uniting two equal parts of ice cream and whipped cream. Such a dessert's Spanish counterpart is called semifrío. In Italian cuisine, the Semifreddo is commonly made with gelato as a primary ingredient. It is typical of the Italian region of Emilia-Romagna.

Sorbet

Sorbet is a frozen dessert made from sweetened water flavoured with fruit (typically juice or puree), wine, and/or liqueur. The origin of sorbet is variously explained as either a Roman invention, or a Middle Eastern drink charbet, made of sweetened fruit juice and water. The term sherbet or charbet is derived from Turkish: þerbat/ °erbet, "sorbet", from the Persian sharbat. Sorbet is sometimes served between courses as a way to cleanse the palate before the main course.

Classification and Description

Sorbet is often confused with Italian ice and often taken to be the same as sherbet. Sorbets/sherbets may also contain alcohol, which lowers the cold temperature, resulting in softer texture. In the UK, *sherbet* refers to a fizzy powder, and only the term *sorbet* would be used. Whereas ice cream is based on dairy products with air copiously whipped in, sorbet has neither, which makes for a dense and extremely flavorful product. Sorbet is served as a non-fat or low-fat alternative to ice cream. In Italy a similar though crunchier textured dish called granita is made. As the liquid in granita freezes it forms noticeably large-size crystals, which are left unstirred. Granita is also often sharded with a fork to give an even crunchier texture when served. Agraz is a type of sorbet, usually associated with the Maghreb and north Africa. It is made from almonds, verjuice, and sugar. It has a strongly acidic flavour, because of the verjuice. (*Larousse Gastronomique*)

Early History and Folklore

One account says that Marco Polo brought a recipe for a sorbet-like dessert on his way back to Italy from China in the late 13th century, as written in an account of his journey, *The Travels of Marco Polo*.

Other folklore holds that Nero, the Roman Emperor, invented sorbet during the first century A.D. when he had runners along the Appian way pass buckets of snow hand over hand from the mountains to his banquet hall where it was then mixed with honey and wine.

Frozen desserts are believed to have been brought to France in 1533 by Catherine de' Medici when she left Italy to marry the Duke of Orleans, who later became Henry II of France. By the end of the 17th century, sorbet was served in the streets of Paris, and spread to England and the rest of Europe.

Distinction from Sherbet

American Terminology

In the United States, sorbet and sherbet (although it's spelled 'sherbet' it is widely pronounced 'sherbert') are distinctly different products. For Americans, sherbet is the more widely-known term and typically designates a fruity flavoured frozen dairy product with a milkfat content between 1 and 2%. Sorbet, on the other hand, is considered by Americans to be a fruity frozen product with no dairy content, similar to Italian ice. Sherbet in the United States must include dairy ingredients such as milk or cream to reach a milkfat content between 1% and 2%. Products with higher milkfat content are defined as ice cream; products with lower milkfat content are defined as water ice.

The use of the term "sorbet" is unregulated and is most commonly used with non-dairy, fruit juice "italian ice" products. Although the American legal definitions indicate that the terms "sorbet" and sherbet are interchangeable, actual usage by Americans and the manufacturers of these products bears a clear distinction. A similar situation occurs in the legal definitions by differing international state governments on what is considered beer.

English/French Labeling

On sherbet packages which have both English and French labels, sherbet is translated to *sorbet laitier* which directly translates into English as *dairy sorbet*, differentiating the milk-containing sherbet from milk-less sorbet.

Kulfi

Kulfi or Qulfi is a popular frozen milk-based dessert from India and Pakistan that is often described as "traditional Indian ice-cream". It is popular throughout neighboring countries in South Asia, Burma (Myanmar), and even the Middle East. It has similarities to ice cream (as popularly understood) in appearance and taste, but is denser and creamier. It comes in various flavours, including cream (malai), raspberry, rose, mango, cardamom (*elaichi*), saffron (*kesar* or *zafran*), the more traditional flavours, as well as newer variations like apple, orange, strawberry, peanut, and avocado. Unlike Western ice creams, kulfi is not whipped, resulting in a solid, dense frozen dessert similar to traditional custard based ice-cream. Thus, it is sometimes considered a distinct category of frozen dairy-based dessert. Due to its density, kulfi takes a longer time to melt than Western ice-cream.

History

Just like any other culture exposed to snow and ice, some people living in the Indian subcontinent, especially those living high in the Himalayas, would have stumbled upon the technique of freezing various sweetened liquids, thus turning them into frozen desserts. These privileges were limited to royalty and upper levels of aristocracy in India until modern day refrigeration technology reached South Asia.

Preparation

Kulfi was traditionally prepared by evaporating sweetened and flavoured milk by slow cooking, with almost continuous stirring to keep milk from sticking to the bottom of the vessel where it might burn, till its volume was reduced by a half, thus thickening it, increasing its fat, protein and lactose density. It has a distinctive taste due to caramelization of lactose and sugar during the lengthy cooking process. The semi-condensed mix is then frozen in tight sealed molds (often kulhars with their mouths sealed) that are then submerged in ice mixed with salt to speed up the freezing process. The ice/salt mix, along with its submerged kulfi molds, is placed in earthen pots or matkas that provide insulation from the external heat and slow down the melting of ice. Kulfi prepared in this manner is hence called 'Matka Kulfi'. Kulfi, thus prepared by slow freezing, also renders a unique smooth mouth feel that is devoid of water crystallization.

More recently Kulfi is prepared from evaporated milk, sweetened condensed milk and heavy (double) cream. Then sugar is added and the mixture is further boiled and cornstarch-water paste is added. This paste thickens the mixture, although it is boiled for an additional few minutes. Then flavourings, dried fruits, cardamom, etc. are added. The mixture is then cooled, put in moulds and frozen. If frozen in individual-portion custard bowls for service with a spoon, bowls are removed from the freezer 10–15 minutes before serving to allow for melting at the edges. It is garnished with ground cardamom, saffron, or pistachio nuts. Kulfi is also served with faloodeh (vermicelli noodles made from starch). In some places, people make it at home and make their own flavours. Traditionally in India and Pakistan, kulfi is sold by street vendors called *kulfiwallahs* who keep the kulfi frozen by placing the moulds inside a large earthenware pot called a "matka", filled with ice and salt. It is served on a leaf or frozen onto a stick. It can be garnished with pistachios, cardamom and similar. Often it is served as Falooda Kulfi which is kulfi with rice noodles, rose syrup and other ingredients. Popular flavours include pistachio, mango, vanilla, and rose.

Spaghettieis

Spaghettieis is a German ice cream specialty that looks like a plate of spaghetti. It was invented by Dario Fontanella in the late 1960s in Mannheim, Germany. In the dish, an often light or white colored ice cream is pressed through a modified Spätzle press or potato ricer to make it look like spaghetti. It is then placed over whipped cream and topped with strawberry sauce (to simulate tomato sauce) and either coconut flakes, grated almonds, or white chocolate shavings to represent the parmesan cheese. Although, as of yet, it is not well known to most people outside Europe, it can be found at some Gelaterias and specialty ice cream shops, at special events and at some hotels and restaurants around the world. Spaghettieis is widely recognized across Germany, and it costs around €3,50 a dish.

Granita

Granita (in Italian also granita siciliana) is a semi-frozen dessert made from sugar, water and various flavourings. Originally from Sicily, although available all over Italy (but granita in Sicily is somewhat different from the rest of Italy), it is related to sorbet and italian ice. However, in most of Sicily, it has a coarser, more crystalline texture.

Food writer Jeffrey Steingarten says that "the desired texture seems to vary from city to city" on the island; on the west coast and in Palermo, it is at its chunkiest, and in the east it is nearly as smooth as sorbet. This is largely the result of different freezing techniques: the smoother types are produced in a gelato machine, while the coarser varieties are frozen with only occasional agitation, then scraped or shaved to produce separated crystals.

Ingredients

Common and traditional flavoring ingredients include lemon juice, mandarin oranges, jasmine, coffee, almonds, mint, and when in season wild strawberries and black mulberries. Chocolate granitas have a tradition in the city of Catania and, according to Steingarten, nowhere else in Sicily. The nuances of the Sicilian ingredients are important to the flavour of the finished granita: Sicilian lemons are a less acidic, more floral variety similar to Meyer lemons, while the almonds used contain some number of bitter almonds, crucial to the signature almond flavour.

Serving Conventions

Granita with coffee is very common in the city of Messina, while granita with almonds is popular in the city of Catania. Granita in combination with a yeast pastry called *brioche* is a common breakfast in summer time. (The Sicilian brioche is generally flatter and wider than the French version.) Granita is often found served as a slush-type drink rather than a dessert, in a paper or plastic cup with a plastic lid and a straw.

Frozen Yogurt

Frozen yogurt (also known as frozen yoghurt or froz yog or by the tradenames FroYo and Frogurt) is a frozen dessert made from, or containing yogurt or other dairy products. It is slightly more tart than ice cream, as well as lower in fat (due to the use of milk instead of cream). It differs from ice milk (more recently termed low-fat or light ice cream), which does not include yogurt as an ingredient.

History

Frozen yogurt was introduced in New England in the 1970s as a soft serve dessert by H. P. Hood under the name *Frogurt*. In 1978, Brigham's, a Boston-based ice cream, candy and sandwich chain, developed and introduced the first packaged frozen yogurt under the name *Humphreez Yogart*. It was originally intended as a healthier alternative to ice cream, but consumers complained about the tart taste. Manufacturers began production of a recipe that tasted sweeter, and frozen yogurt took off in the 1980s, reaching sales of $25 million in 1986. In the early 1990s, frozen yogurt was 10% of the frozen dessert market. The first frozen yogurt shop in the UK, Lick, based in Brighton opened in March 2008. Later that year, frozen yogurt shops began opening in London and across the country.

Production

Frozen yogurt usually consists of milk solids, milk fat, yogurt culture, sweetener, corn syrup, colouring, and flavoring.

Milk fat comprises about 0.5-6% of the yogurt. Added in quantities inversely proportional to the amount of milk solids, the milk fat lends richness to the yogurt. Milk solids account for 8-14% of the yogurt's volume, providing lactose for sweetness and proteins for smoothness and increased resistance to melting. Cane or beet sugar provides 15-17% of the yogurt's ingredients. In addition to adding sweetness, the sugar increases the volume of solid ingredients in the yogurt, improving body and texture. Animal gelatins and vegetable additives (guar gum, carrageenan, etc.) stabilize the yogurt, reducing crystallization and increasing the temperature at which the yogurt will melt. This stabilization ensures that the frozen yogurt maintains a smooth consistency regardless of handling or temperature change.

Frozen yogurt can be made in an ice cream machine; however, major companies often use assembly lines specifically dedicated to frozen yogurt production. The milk products and gelatin are combined and homogenized.

They are then cooled to 0 degrees celsius. At this temperature, the yogurt culture is added and the mixture is warmed to 4 degrees Celsius. Once it has reached the desired temperature and viscosity, the yogurt is allowed to sit in aging tanks for up to four hours. Sweeteners, colorings and flavourings are then mixed in, and the yogurt mixture is cooled to -6 to -2 degrees Celsius. To create extra volume and smooth consistency, air is incorporated into the yogurt as the mixture is agitated. When a sufficient amount of air has been incorporated into the product, the yogurt is rapidly frozen to prevent the formation of large ice crystals, and stored in a cold place to be shipped.

Uses

Frozen yogurt has come to be used much like ice cream, and is served in a wide variety of flavours and styles. Many companies allow customers the option of adding various toppings, or ordering their frozen yogurt in cups or in cones. Certain sellers even offer sugar-free varieties. Frozen yogurt made by chains such as Yogurt Fusion, Yogurt Story, Yoforia, Pinkberry, Red Mango, and Yogen Früz is tarter and closer to the original recipe, whereas other companies like TCBY and I Can't Believe It's Yogurt focus on making their frozen yogurt taste like ice cream.

Franchising

It is expected to be a multi-billion dollar industry within the next few years, creating a prime opportunity for franchise expansion. "Which franchise categories holds the most promise for 2009. Frozen yogurt is also turning out to be a simply irresistible opportunity", Entrepreneur Magazine. In 1999, Entrepreneur Magazine rated Yogen Fruz the number 1 franchise in the world that year, ahead of perennial winners such as Subway and McDonald's. Popular national frozen yogurt franchises include Yogen Früz, Pinkberry, Menchie's, Red Mango, TCBY, Cherry on Top and Tutti Frutti.

Ghee

Ghee is a class of clarified butter that originated in South Asia, and is commonly used in South Asian (Indian, Bangladeshi, Nepali and Pakistani), North African (Egyptian and Berber) and Horn African cuisine (Somali, Ethiopian and Eritrean).

Preparation

Ghee, also known as *clarified butter* in anglophone countries, is made by simmering unsalted butter in a cooking vessel until all water has boiled off, the milk solids (or protein) have settled to the bottom, and a scum has floated on top. After removing the scum, the cooked and clarified butter is then spooned off or tipped out carefully to avoid disturbing the milk solids on the bottom of the pan. Ghee can be stored for extended periods without refrigeration, provided that it is kept in an airtight container to prevent oxidation and remains moisture-free. The texture, colour, or taste of ghee depends on the source of the milk from which the butter was made and the extent of boiling and simmering.

Religious Use

Ghee made from cow's milk has a sacred role in Vedic and modern Hindu libation and anointment rituals. There is also a hymn to ghee. Ghee is also burnt in the Hindu religious ritual of *Ârati* (Aarti) and is the principal fuel used for the Hindu votive lamp known as the *diyâ* or *dîpa* (deep). It is used in marriages and funerals, and for bathing *mûrtis* (divine idols) during worship.

In other religious observances, such as the prayers to *Œiva* (Shiva) on *Mahâ-úivarâtrî* (Maha Shivaratri), ghee is served in *Pañcâm[ta* (Panchamruta) along with four other sacred substances: sugar, honey, milk, and *dahî* (yogurt). According to the *Mahâbhârata*, ghee is the very root of sacrifice by Bhîcma. Also, it is used generously in *homam* or *yajña* since it is considered as food for the Devas.

Usage in Food

Ghee is widely used in Indian cuisine and Pakistani cuisine. It, however, is mentioned in the Epic of Gilgamesh, and is probably Akkadian in origin. In many parts of India and Pakistan, especially in Punjab, Haryana, Gujarat, Maharashtra, Bengal, Orissa and many other states, rice is traditionally prepared or served with ghee (including biryani). In the Bharuch district of Gujarat, Ghee is served with kichri, usually an evening meal of yellow rice with curry, a sauce made from yoghurt, cumin seeds, curry leaves, ghee, cornflour, tumeric, garlic and salt. Ghee is also an ingredient as well as used in the preparation of *kadhi* and used in Indian and Pakistani sweets such as Mysore pak , and different varieties of halva and laddu. Punjabi cuisine prepared in restaurants uses large amounts of ghee. Naan and roti are sometimes brushed with ghee, either during preparation or while serving. Ghee is an ideal fat for deep frying because its smoke point (where its molecules begin to break down) is 250°C (485°F), which is well above typical cooking temperatures of around 200°C (400°F) and above that of most vegetable oils.

Nutrition

Like any clarified butter, ghee is composed almost entirely of fat; the nutrition facts label found on bottled cow's ghee produced in the USA indicates eight mg. of cholesterol per teaspoon.

Ghee has been shown to slightly, but not significantly, reduce serum cholesterol in one rodent study. Studies in Wistar rats have revealed one mechanism by which ghee reduces plasma LDL cholesterol. This action is mediated by an increased secretion of biliary lipids. Indian restaurants and some households may use hydrogenated vegetable oil (also known as *vanaspati, dalda,* or "vegetable ghee") in place of ghee due to its lower cost. This "vegetable ghee" may contain trans fat. Trans fats are increasingly linked to serious chronic health conditions. The term *shuddh ghee,* however, is not officially enforced in many regions, so partially hydrogenated oils are marketed as pure ghee in some areas. Where this is illegal in India, law enforcement often cracks down on the sale of fake ghee. Ghee is also sometimes called *desi* (country-made) ghee or *asli* (genuine) ghee to distinguish it from "vegetable ghee".

Outside South Asia

Several cultures make ghee outside of South Asia. Egyptians make a product called *samna baladi,* literally meaning "local ghee"; i.e., Egyptian ghee virtually identical to ghee in terms of process and end result. In Ethiopia, [[*niter kibbeh*]] is made and used in much the same way as ghee, but with spices added during the process that result in a distinctive taste.

Moroccans (especially Berbers) take this one step further, aging spiced ghee in the ground for months or even years, resulting in a product called *smen.* In northeastern Brazil, an unrefrigerated butter very similar to ghee, called *manteiga-de-garrafa* (butter-in-a-bottle) or *manteiga-da-terra* (butter of the land), is common. In Europe, it is also widely used. For example, *Wiener Schnitzel* is traditionally fried in a version of ghee called *Butterschmalz.*

Whey

Whey or milk plasma is the liquid remaining after milk has been curdled and strained. It is a by-product of the manufacture of cheese or casein and has several commercial uses. *Sweet whey* is manufactured during the making of rennet types of hard cheese like Cheddar or Swiss.

Acid whey (also known as "sour whey") is obtained during the making of acid types of cheese such as cottage cheese.

Production

Whey is a co-product of cheese production. It is one of the components that separate from milk after curdling, when rennet or an edible acidic substance is added.

Uses

Whey is used to produce ricotta, brown cheeses, Messmör/Prim, and many other products for human consumption. It is also an additive in many processed foods, including breads, crackers, and commercial pastry, and in animal feed. Whey proteins consist primarily of á-lactalbumin and â-lactoglobulin. Depending on the method of manufacture, whey may also contain glycomacropeptides (GMP).

Whey protein (derived from whey) is often sold as a nutritional supplement. Such supplements are especially popular in the sport of bodybuilding. In Switzerland, where cheese production is an important industry, whey is used as the basis for a carbonated soft drink called Rivella.

In Iceland, MS manufactures and sells liquid whey as *Mysa* in 1-liter cartons (energy 78 kJ or 18 kcal, calcium 121 mg, protein 0.4 g, carbohydrates 4.2 g, sodium 55 mg).

Throughout history, whey was a popular drink in inns and coffee houses. When Joseph Priestley was at college at Daventry Academy 1752–1755, he records that, during the morning of Wednesday 22 May 1754, he "went with a large company to drink whey."

This was probably 'sack whey' or 'wine whey.' A contemporary recipe for 'wine whey' instructs: "Put a pint of skimmed milk, and half a pint of white wine into a bason, let it stand a few minutes, then pour over it a pint of boiling water, let it stand a little, and the curd will gather in a lump, and settle to the bottom, then pour your whey into a China bowl, and put in a lump of sugar, a sprig of balm, or a slice of lemon."

An alternative recipe is for 'Cream of Tartar Whey': "Put a pint of blue milk over the fire, when it begins to boil, put in two tea spoonfuls of cream of tartar, then take it off the fire, and let it stand till the curd settles to the bottom of the pan, then put it into a bason to cool, and drink it milk warm." Whey was also used in central Spain to enrich bakery products. In some traditions, it was used instead of water to produce bread dough.

Whey Cream and Butter

Cream can be skimmed from whey. Whey cream is more salty, tangy, and "cheesy" than ("sweet") cream skimmed from milk, and can be used to make whey butter. Whey cream and butter are suitable for making butter-flavoured food, as they have a stronger flavour of their own. They are also cheaper than sweet cream and butter.

Health

Because whey contains lactose, it should be avoided by those who are lactose intolerant. Dried whey, a very common food additive, contains more than 70% lactose. When used as a food additive, whey can contribute to quantities of lactose far above the level of tolerance of most lactose-intolerant individuals.

Liquid whey contains lactose, vitamins, protein, and minerals, along with traces of fat. In 2005, researchers at Lund University in Sweden discovered that whey appears to stimulate insulin release, in type 2 diabetics. Writing in the *American Journal of Clinical Nutrition*, they also discovered that whey supplements can help regulate and reduce spikes in blood sugar levels among people with type 2 diabetes by increasing insulin secretion.

Protein

Whey protein is the name of globular proteins that can be isolated from whey. It is typically a mixture of globinstagers beta-lactoglobulin (~65%), alpha-lactalbumin (~25%), and serum albumin (~8%), which are soluble in their native culture forms, independent of pH.

Casein and Caseinates

Casein (from Latin *caseus*, "cheese") is the name for a family of related Phosphoprotein proteins. These proteins are commonly found in mammalian milk, making up 80% of the proteins in cow milk and between 20% to 45% of the proteins in human milk.

Casein has a wide variety of uses, from being a major component of cheese, to use as a food additive, to a binder for safety matches. As a food source, casein supplies essential amino acids as well as some carbohydrates and the inorganic elements calcium and phosphorus.

Description

Casein consists of a fairly high number of proline peptides, which do not interact. There are also no disulfide bridges. As a result, it has relatively little tertiary structure. Because of this, it cannot denature. It is relatively hydrophobic, making it poorly soluble in water. It is found in milk as a suspension of particles called casein micelles which show some resemblance with surfactant-type micellae in a sense that the hydrophilic parts reside at the surface. The caseins in the micelles are held together by calcium ions and hydrophobic interactions. There are several models that account for the special conformation of casein in the micelles. One of them proposes that the micellar nucleus is formed by several submicelles, the periphery consisting of microvellosities of ê-casein. Another model suggests that the nucleus is formed by casein-interlinked fibrils (Holt, 1992). Finally, the most recent model (Horne, 1998) proposes a double link among the caseins for gelling to take place. All 3 models consider micelles as colloidal particles formed by casein aggregates wrapped up in soluble ê-casein molecules. The isoelectric point of casein is 4.6. Since milk's pH is 6.6, casein has a negative charge in milk. The purified protein is water insoluble. While it is also insoluble in neutral salt solutions, it is readily dispersible in dilute alkalis and in salt solutions such as sodium oxalate and sodium acetate. The enzyme trypsin can hydrolyze off a phosphate-containing peptone. It is used to form a type of organic adhesive.

Uses

Paint

Casein paint, is a fast-drying, water-soluble medium used by artists. Casein paint has been used since ancient Egyptian times as a form of tempera paint, and was widely used by commercial illustrators as the material of choice until the late 1960s when, with the advent of acrylic paint, casein became less popular.

Glue

Casein-based glues were popular for woodworking, including for aircraft as late as the de Havilland Mosquito.

Cheese making

Cheese consists of proteins and fat from milk, usually the milk of cows, buffalo, goats, or sheep. It is produced by coagulation of casein. Typically, the milk is acidified and then coagulated by the addition of rennet, a proteolytic enzyme typically obtained from the stomachs of calves. The solids are separated and pressed into final form. Unlike many proteins, casein is not coagulated by heat. During the process of clotting, milk-clotting proteases act on the soluble portion of the caseins, K-Casein, thus originating an unstable micellar state that results in clot formation. When coagulated with chymosin, casein is sometimes called paracasein. Chymosin (EC 3.4.23.4) is an aspartic protease that specifically hydrolyzes the peptide bond in Phe105-Met106 of ê-casein and is considered to be the most efficient protease for the cheese-making industry (Rao et al., 1998). British terminology, on the other hand, uses the term caseinogen for the uncoagulated protein and casein for the coagulated protein. As it exists in milk, it is a salt of calcium.

Plastics and Fiber

Some of the earliest plastics were based on casein. In particular Galalith was well-known for use in buttons. Fiber can be made from extruded casein.

Protein Supplements

An attractive property of the casein molecule is its ability to form a gel or clot in the stomach. The ability to form this clot makes it very efficient in nutrient supply. The clot is able to provide a sustained slow release of amino acids into the blood stream, sometimes lasting for several hours. This provides better nitrogen retention and utilization by the body. Plasma immunoreactive IGF-1 concentration in rats given a casein diet was higher than that in rats given a soya-bean-protein or protein-free diet.

Medical and Dental Uses

Casein-derived compounds are used in tooth remineralization products to stabilize amorphous calcium phosphate (ACP) and release the ACP onto tooth surfaces where it can facilitate remineralization.

Controversies

Autism

Casein has been documented to break down to produce the peptide casomorphin, an opioid that appears to act primarily as a histamine releaser. Some research indicates that this casomorphine aggravates the symptoms of autism. A 2006 review of seven studies indicated that although benefits were seen in all studies from the introduction of elimination diets (e.g., casein or gluten free) in the treatment of autism spectrum disorders, none of the studies were performed in a manner to create an unbiased scientific opinion. Preliminary data from the first and only double-blind randomized control trial of a gluten- and casein-free diet "indicated no statistically significant findings even though several parents reported improvement in their children." Although research has shown high rates of use of complementary and alternative therapies (CAM) for children with autism including gluten and/or casein exclusion diets, the evidence for efficacy of these diets is currently unsubstantiated.

A1/A2 Beta Caseins

Four casein proteins make up about 80% of the protein in cow's milk. One of the major caseins is beta-casein, of which there are several types, but "A1" and "A2" are the most common. According to the Australian Food Standards Agency. Certain breeds of cows, such as Friesians, produce mostly A1 milk, whereas other breeds, such as Guernseys, as well as sheep and goats, produce mostly A2 milk. Food Standards Australia New Zealand also reports that despite some hypotheses that consumption of A2 milk "provides levels of protection to consumers from autism in children as well as schizophrenia, diabetes and heart disease," the scientific evidence for such claims is "very limited". Additionally, the European Food Safety Authority carried out a literature review in 2009 concluding that "a cause and effect relationship is not established between the dietary intake of BCM7, related peptides or their possible protein precursors and non-communicable diseases". Studies supporting these claims have had significant flaws, and the data are inadequate to guide autism treatment recommendations.

T. Colin Campbell's *The China Study* (2006), a univariate correlative published book, suggests a correlation between powdered, isolated casein administered to rats and the promotion of cancer cell growth when exposed to carcinogens. A 2001 study suggests that another milk protein, whey protein, but not casein, may play a protective role against colon tumors in rats. According to a study from the Australian Dairy Council, casein has antimutagenic effects.

Casein-free Diet

Casein has a molecular structure that is quite similar to that of gluten. Thus, some gluten-free diets are combined with casein-free diets and referred to as a gluten-free, casein-free diet. Casein is often listed as sodium caseinate, calcium caseinate or milk protein. These are often found in energy bars, drinks, and packaged goods. A small fraction of the population is allergic to casein.

Altering the Effects of Polyphenols

A study of Charité Hospital in Berlin by Lorenzo et al., published in The European Heart Journal, showed that adding milk to tea causes the casein to bind to the molecules in tea that cause the arteries to relax, especially a catechin molecule called EGCG, although a more recent study by Reddy et al. (2005) suggests that the addition of milk to tea does not alter the antioxidant activity in vivo and the cardiovascular effect remains controversial. A study published in the journal *Free Radical Biology and Medicine* indicates that casein reduced the peak plasma levels of beneficial polyphenols after the consumption of blueberries.

Casein Products

Casein is the principal protein found in cow.s milk from which it has been extracted commercially for most of the 20th century. It is responsible for the white, opaque appearance of milk in which it is combined with calcium and phosphorus as clusters of casein molecules, called micelles.

The major uses of casein until the 1960s were in technical, non-food applications such as adhesives for wood, in paper coating, leather finishing and in synthetic fibres, as well as plastics for buttons, buckles *etc*. During the past 30 years, however, the principal use of casein products has been as an ingredient in foods to enhance their physical (so-called .functional.) properties, such as whipping and foaming, water binding and thickening, emulsification and texture, and to improve their nutrition.

In New Zealand, casein is precipitated from the skim milk that is produced after centrifugal separation of whole milk. The skim milk may be acidified to produce acid casein or treated with an enzyme, resulting in the so-called rennet casein. The precipitated casein curd is separated from the whey, washed and dried. Water-soluble derivatives of acid caseins, produced by reaction with alkalis, are called caseinates.

The amount of casein in cow.s whole milk varies according to the breed of cow and stage of lactation, but is generally in the range 24-29 g L-1. Casein contains 0.7-0.9% phosphorus, covalently bound to the protein by a serine ester linkage. Casein is consequently known as a phospho-protein. All the amino acids that are essential to man are present in casein in high proportions, with the possible exception of cysteine. Thus, casein may be considered as a highly nutritious protein.

Casein exists in milk in complex groups of molecules (sometimes referred to as calcium phospho-caseinate) that are called .micelles.. The micelles consist of casein molecules, calcium, inorganic phosphate and citrate ions, and have a typical molecular weight of several hundred million. In terms of physical chemistry, the casein micelles may be considered to exist in milk as a very stable colloidal dispersion.

The caseins, as proteins, are made up of many hundreds of individual amino acids, each of which may have a positive or a negative charge, depending on the pH of the [milk] system. At some pH value, all the positive charges and all the negative charges on the [casein] protein will be in balance, so that the net charge on the protein will be zero. That pH value is known as the isoelectric point (IEP) of the protein and is generally the pH at which the protein is least soluble.

For casein, the IEP is approximately 4.6 and it is the pH value at which acid casein is precipitated. In milk, which has a pH of about 6.6, the casein micelles have a net negative charge and are quite stable.

Although casein has been shown to consist of several individual casein components, referred to as ás1-, ás2-, â- and ê-casein, each having slightly different properties (which are caused by small variations in their amino acid content).

Extraction of Casein from Milk

Separation

Whole cow.s milk (with a typical fat content of 4.65%) is first separated by means of centrifuges that produce cream (for the manufacture of butter or other milkfat products) and skim milk. Skim milk can thus be considered as the raw material from which casein products are made.

Precipitation

Precipitation by means of acidification can be considered in terms of simple chemistry as follows, R being the casein protein:

H2N-R-COO- + H+ → +H3N-R-COOcasein micelle acid casein (pH = 6.6) (pH = 4.6)

Colloidal Dispersion Insoluble Particles

In the case of enzyme coagulation of casein, there is no change in the pH of the milk. Instead, the addition of a specific enzyme, *chymosin*, which is found in the stomach of newborn calves, specifically cleaves a highly charged portion from the ê-casein, called glycomacropeptide. That action causes the remainder of the ê-casein (now called *para*-ê-casein) to lose its considerable power in stabilising the micelles in milk, and the result is the formation of a three-dimensional gel network or clot of the casein in the presence of calciumions. This reaction is essential in the manufacture of virtually all cheese types and in the production of rennet casein.

Wet-processing Operations

When the casein has been precipitated, the mixture is heated (a process known in the dairy industry as .cooking.). Heating of the precipitated casein causes the particles to shrink and expel moisture (whey) (rather like a sponge), and also to agglomerate together to form clumps of curd. The curd is then separated from the whey and washed several times with water in vats prior to mechanical dewatering by pressing or centrifuging.

Drying and Dry Processing of Casein

The dewatered curd, with a moisture content of about 55%, is dried by means of hot air using either fluidised bed driers with multiple decks or pneumatic-conveying ring driers to produce a dry casein having a moisture content of 10-12%. The warm, unmilled casein is then subjected to several dry processing steps, including cooling (usually by air conveying), .tempering. or conditioning to ensure that moisture is distributed evenly between large and small particles, milling, sifting (to produce coarse, medium and fine mesh particles), blending (to ensure uniformity) and bagging. The 25 kg bags of casein are placed on pallets and stored ready for shipping.

Types of Casein

As indicated above, two basic types of casein - acid and rennet - are produced in New Zealand. They are named in accordance with the coagulating agent employed. Three types of acid casein are made commercially: lactic, hydrochloric and sulphuric acid caseins. In New Zealand, lactic acid casein has been the most common casein product, although larger quantities of sulphuric acid casein have been produced in recent years. In Australia and Europe, the most common precipitant for acid casein is hydrochloric acid, which is a by-product of the chemical industry and hence is relatively cheap. In New Zealand, however, which has a very small chemical industry, hydrochloric acid is relatively expensive.

On the other hand, sulphuric acid is relatively cheap, being produced in comparatively large quantities by the fertiliser industry for use in the manufacture of superphosphate. Consequently, virtually all mineral acid casein made in New Zealand is precipitated using sulphuric acid.

The properties of the different types of acid casein are very similar and, for most applications, the acid caseins can be used interchangeably. For the manufacture of rennet casein, several different coagulants are now available. These include chymosin (previously known as .rennet. or .rennet extract.), the milk-clotting enzyme extracted from the stomachs of young calves, and a number of so-called microbial rennets, which are enzymes that have been produced by means of microbial fermentation techniques. The caseins produced using any of these enzyme preparations are all known as rennet casein, and all have similar properties. However, their properties are noticeably different from those of acid casein.

Acid Casein Manufacture

Lactic Acid Casein

For the manufacture of lactic acid casein, skim milk (pH 6.6) is first pasteurised (72°C for 15 s). It is then cooled to setting temperature (22-26°C) and inoculated with several strains of lactic acid-producing bacteria, known as .starters. (*e.g. Lactococcus lactis* sub-species *cremoris*, 0.1-0.2% of milk volume). The milk is incubated, without agitation, in large silos (each with a capacity of up to 250 000 L) for a period of 14-16 h. During this period, some of the lactose in the milk is converted to lactic acid by the starter and the pH is reduced to about 4.6, causing coagulation of the casein (and the milk). This takes the form of a soft gel and is generally referred to in the industry as .coagulum. or .coag.

The fermentation of lactose to lactic acid in the manufacture of lactic acid casein is not as simple. however, and a number of other compounds are produced as well, *e.g.* diacetly ($CH_3COCOCH_3$), acetoin ($CH_3CHOHCOCH_3$) and benzoin ($C_6H_5COCHOHC_6H_5$).

These are present in relatively small amounts and do not generally present any serious problems. After the pH of the milk has reached 4.6-4.7, the coagulum is .cooked. (*i.e.* heated), usually by means of a combination of indirect heating (through a heat exchanger) and steam injection, to a temperature of 50-55oC. After a brief period of residence in a cooking line and acidulation.

Mineral Acid Casein

For the precipitation of mineral acid casein, pasteurised skim milk at a pH of 6.6 is mixed thoroughly with dilute (0.25 mol L-1) acid at a temperature of about 20°C to a pH of approximately 4.6. In this case, because of the very vigorous agitation and the short mixing time, the casein is precipitated as very fine, individual particles in a liquid serum (whey), unlike the gel/coagulum formed in lactic acid casein manufacture. The acidified milk mixture is then cooked and processed further in a manner similar to that described for the production of lactic acid casein.

Rennet Casein Manufacture

When rennet casein is made, the skim milk is not acidified. Hence, the pH remains at 6.6 throughout the manufacturing process. Following pasteurisation of the skim milk, it is cooled to a setting temperature of about 29°C, and calf rennet or microbial rennet is added (ratio by volume: 1 of rennet to about 7000 of skim milk) and mixed in thoroughly. During the first stage of renneting, the enzyme specifically cleaves one of the bonds in ê-casein, releasing part of the protein chain that is commonly referred to as glycomacropeptide. This action destabilises the casein micelles and, in the second stage of the reaction, they form a three-dimensional clot with some of the calcium ions that are present in the milk. The renneting process usually takes place in a period of 20-40 min under the conditions (pH and temperature) described above. The clotted milk may then be cooked and the casein processed in a manner similar to that described for lactic acid casein.

Yield

The yield of commercial casein is about 3 kg/100 kg skim milk.

Manufacture of Caseinates

Caseinates may be produced by reaction under aqueous conditions of acid casein curd or dry acid casein with any one of several different dilute alkalis. The resulting homogeneous solution may be spray dried to produce a caseinate powder having a moisture content of 3-6%, depending on the manufacturing conditions and customer requirements.

Spray-dried Sodium Caseinate

The most common alkali used in the manufacture of spray dried sodium caseinate is sodium hydroxide. It is mixed (as an aqueous solution with a typical concentration of 2.5 M) with a slurry of the casein curd or powder in water. The usual amount of sodium hydroxide needed is about 2% (w/w) of the casein solids.

The casein curd is milled using one or more colloid mills to reduce the size of the individual particles so that they will dissolve rapidly, and is then mixed with the alkali using high shear. In producing solutions of sodium caseinate for spray drying, it is important to achieve the maximum possible concentration of solids for economic reasons, as the more water there is to evaporate, the higher is the energy cost. Concentrated solutions of sodium caseinate (> 15% solids), however, are very viscous, and require powerful agitators and pumps for mixing and fluid transfer.

The use of high temperatures (60-95°C) during the later dissolving stages is of practical benefit, as the viscosity of sodium caseinate solutions decreases with temperature. However, there is still a delicate balance between what is mechanically achievable and what is economically practicable during the commercial production of sodium caseinate.

Other Caseinates

The manufacture of potassium and ammonium caseinates is very similar to that of sodium caseinate, although, in the case of ammonium caseinate, a lot of the ammonia is evaporated from the solution during the drying process.

A solution of sodium caseinate, like those of potassium and ammonium caseinates, has a straw-like colour and is completely different in appearance from milk. Solutions of calcium caseinate, on the other hand, are very white and opaque - even whiter than milk, and they are less viscous than solutions of the other caseinates. Calcium caseinate solutions are produced by adding a slurry of lime (calcium hydroxide) in water to a casein curd-water mixture and reacting the combined slurries at a relatively low temperature (< 45°C) until the neutralisation reaction is completed. Use of higher temperatures before neutralisation is completed is likely to result in precipitation or coagulation of the partly reacted calcium caseinate, with probable dumping of the contents of the reaction vessel. All caseinate powders have a white appearance.

Composition of Casein Products

As noted earlier, when the same manufacturing operations are used, the caseins produced from lactic, sulphuric or hydrochloric acid precipitation are almost indistinguishable from one another.

During the acidification process in the manufacture of acid casein, the calcium and inorganic phosphate (that are associated with the casein micelle in milk) are dissolved and leached from the curd leaving only the organic phosphorus and a small residue of calcium. Rennet casein contains about 3% calcium and approximately 1.4% phosphorus.

As they are spray-dried products, their moisture content is much lower than that of the caseins, and their protein content is consequently higher. With a pH generally in the range 6.5-7.0, sodium caseinate will usually contain 1.2-1.4% sodium, whereas the calcium content of calcium caseinate is generally in the range 1.3-1.6%.

Properties of Casein Products

Solubility

Acid and rennet casein are insoluble in water. Virtually all applications of casein products require them to be dissolved first. Consequently, before use, acid casein must be dissolved using an alkali to produce a solution with a pH of 6.5 or higher. The caseinates mentioned in the previous sections are used for food and pharmaceutical applications. For non-food, technical applications, acid casein may be dissolved in other alkalis such as borax or ammonia, usually to a somewhat higher pH (7.5-9.5, or higher) than that used for edible applications.

Water Absorption and Viscosity

Casein products can absorb substantial amounts of water, so they can modify the texture of dough or baked products, serve as the matrix former in cheese-type products, produce specialised plastic materials, or increase the consistency of solutions such as soups. They are good film-formers and find use in whipping and foaming applications, and in emulsions of fats or oils in water.

Nutrition

The nutritional quality of a protein is determined primarily by its essential amino acid content. For adult man, eight amino acids are essential, *i.e.* they must be supplied in the diet. These are isoleucine, leucine, lysine, methionine, phenylalanine, threonine, tryptophan and

valine; the infant requires histidine as well. In comparison with an
.ideal. reference protein composition that was developed by the FAO
in 1973, casein contains an adequate supply of all the essential amino
acids with the possible exception of the sulphur-containing amino
acids methionine and cysteine.

Curd

Curds are a dairy product obtained by *curdling* (coagulating) milk
with rennet or an edible acidic substance such as lemon juice or
vinegar, and then draining off the liquid portion (called whey). Milk
that has been left to sour (raw milk alone or pasteurized milk with
added lactic acid bacteria or yeast) will also naturally produce curds,
and sour milk cheese is produced this way. The increased acidity
causes the milk proteins (casein) to tangle into solid masses, or *curds*.
The rest, which contains only whey proteins, is the whey. In cow's
milk, 80% of the proteins are caseins.

Curd products vary by region and include cottage cheese, quark
(both curdled by bacteria and sometimes also rennet) and paneer
(curdled with lemon juice). The word can also refer to a non-dairy
substance of similar appearance or consistency, though in these cases
a modifier or the word *curdled* is generally used (e.g., bean curds,
lemon curd, or curdled eggs). In England, curds produced from the
use of rennet is referred to as junket, with true curds and whey only
occurring from the natural separation of milk due to its environment
(temperature, acidity).

In Asia, curds are essentially a vegetarian preparation using
yeast to ferment the milk. In some places in Indian subcontinent,
particularly in North India, buffalo milk is used for curd due to its
higher fat content, making a thicker curd. The quality of curds depends
on the starter used. The time taken to curdle also varies with the
seasons, taking less than 6 hours in hot weather and up to 16 hours
in cold weather. In the industry, an optimal temperature of 43 °C for
4–6 hours is used for preparation.

In India, the word *curd* means only plain yogurt. In South India,
it is common practice to finish any meal with yogurt or buttermilk.
In Sweden, curds is a major ingredient of the traditional baked
cheesecakes *Ostkaka* (with egg, flour, almonds), which is eaten with
jam and cream.

In Turkey, curds call as ke° and very common in breakfast onto
fried bread and also is eaten into macaroni in the region of Bolu and

Zonguldak provinces. Cheese curds, drained of the whey and served without further processing or aging, are popular in some French-speaking regions of Canada, such as Quebec and parts of Ontario. The image to the right shows freshly made morsels of Cheddar cheese before being pressed and aged. In Quebec and Eastern Ontario, cheese curds are popularly served with french fries and gravy as *poutine*. In some parts of the U.S., they are breaded and fried, or are eaten straight. Cheese curds may also be coated with a powdered flavoring agent and sold as a snack food in a fashion similar to flavoured potato chips.

Yoghurt

Yoghurt, yogurt or yogourt (Turkish: *Yoðurt*) is a dairy product produced by bacterial fermentation of milk. The bacteria used to make yoghurt are known as "yoghurt cultures". Fermentation of lactose by these bacteria produces lactic acid, which acts on milk protein to give yoghurt its texture and its characteristic tang.

Worldwide, cow milk is most commonly used to make yoghurt, but milk from water buffalo, goats, sheep, camels and yaks is also used in various different parts of the world. In theory the milk of any mammal could be used to make yoghurt. Soya yoghurt, a non-dairy yoghurt alternative, is made from soy milk; this is not an animal product, being made from soy beans. Dairy yoghurt is produced using a culture of *Lactobacillus delbrueckii* subsp. *bulgaricus* and *Streptococcus salivarius* subsp. *thermophilus* bacteria. In addition, *Lactobacillus acidophilus*, *Lactobacillus bifidus* and *Lactobacillus casei* are also sometimes used in culturing yoghurt. The milk is first heated to about 80 °C to kill any undesirable bacteria and to denature the milk proteins so that they set together rather than form curds. The milk is then cooled to about 45 °C. The bacteria culture is added, and the temperature is maintained for 4 to 7 hours to allow fermentation.

Etymology and Spelling

The word is derived from Turkish: *yoðurt*, and is related to the obsolete verb *yoðmak* 'to be curdled or coagulated; to thicken'. The letter ð was traditionally rendered as "gh" in transliterations of Turkish. In older Turkish, the letter denoted a voiced velar fricative /c/, but this sound is elided between back vowels in modern Turkish, in which the word is pronounced. Some eastern dialects retain the consonant in this position, and Turks in the Balkans pronounce the word with a hard /a/.

In English, there are several variations of the spelling of the word. In Australia and New Zealand "yoghurt" prevails. In the United Kingdom "yoghurt" and "yogurt" are both current, "yogurt" being more common on product labels, and "yoghourt" is an uncommon alternative. In the United States, "yogurt'" is the usual spelling and "yoghurt" a minor variant. In Canada, "yogurt" is most common among English speakers, but many brands use "yogourt," since it is an acceptable spelling in both official languages. Whatever the spelling, the word is usually pronounced with a short *o* in the UK, with a long *o* in North America, Australia and South Africa, and with either a long or short *o* in New Zealand and Ireland.

History

There is evidence of cultured milk products in cultures as far back as 2000 BC. The earliest yoghurt was probably fermented spontaneously, perhaps by wild bacteria residing inside goatskin bags used for transportation. In the early 1800s, men used yogurt to clean their goats and sheep. Many women also used yogurt to wash their bodies and hair.

Yogurt was the best known cleaning agent at the time. The oldest writings mentioning yoghurt are attributed to Pliny the Elder, who remarked that certain nomadic tribes knew how "to thicken the milk into a substance with an agreeable acidity". The use of yoghurt by medieval Turks is recorded in the books *Diwan Lughat al-Turk* by Mahmud Kashgari and *Kutadgu Bilig* by Yusuf Has Hajib written in the 11th century. Both texts mention the word "yoghurt" in different sections and describe its use by nomadic Turks. An early account of a European encounter with yoghurt occurs in French clinical history: Francis I suffered from a severe diarrhoea which no French doctor could cure. His ally Suleiman the Magnificent sent a doctor, who allegedly cured the patient with yoghurt. Being grateful, the French king spread around the information about the food which had cured him.

Until the 1900s, yoghurt was a staple in diets of people in the Russian Empire (and especially Central Asia and the Caucasus), Western Asia, South Eastern Europe/Balkans, Central Europe, and India. Stamen Grigorov (1878–1945), a Bulgarian student of medicine in Geneva, first examined the microflora of the Bulgarian yoghurt. In 1905, he described it as consisting of a spherical and a rod-like lactic acid bacteria. In 1907 the rod-like bacteria was called *Lactobacillus bulgaricus* (now *Lactobacillus delbrueckii subsp.*

bulgaricus). The Russian Nobel laureate biologist Ilya Ilyich Mechnikov, from the Institut Pasteur in Paris, was influenced by Grigorov's work and hypothesised that regular consumption of yoghurt was responsible for the unusually long lifespans of Bulgarian peasants.

Believing *Lactobacillus* to be essential for good health, Mechnikov worked to popularise yoghurt as a foodstuff throughout Europe. Isaac Carasso industrialised the production of yoghurt. In 1919, Carasso, who was from Ottoman Salonika, started a small yoghurt business in Barcelona and named the business Danone ("little Daniel") after his son. The brand later expanded to the United States under an Americanised version of the name: Dannon. Yoghurt with added fruit jam was patented in 1933 by the Radlická Mlékárna dairy in Prague. It was introduced to the United States in 1947, by Dannon.

Yoghurt was first introduced to the United States in the first decade of the twentieth century, influenced by Élie Metchnikoff's *The Prolongation of Life; Optimistic Studies* (1908); it was available in tablet form for those with digestive intolerance and for home culturing. It was popularised by John Harvey Kellogg at the Battle Creek Sanitarium, where it was used both orally and in enemas, and later by Armenian immigrants Sarkis and Rose Colombosian, who started "Colombo and Sons Creamery" in Andover, Massachusetts in 1929. Colombo Yoghurt was originally delivered around New England in a horse-drawn wagon inscribed with the Armenian word "madzoon" which was later changed to "yogurt", the Turkish name of the product, as Turkish was the lingua franca between immigrants of the various Near Eastern ethnicities who were the main consumers at that time. Yoghurt's popularity in the United States was enhanced in the 1950s and 1960s, when it was presented as a health food. By the late 20th century yoghurt had become a common American food item and Colombo Yogurt was sold in 1993 to General Mills, which discontinued the brand in 2010.

Nutritional Value and Health Benefits

Yoghurt is nutritionally rich in protein, calcium, riboflavin, vitamin B6 and vitamin B12. It has nutritional benefits beyond those of milk. People who are moderately lactose-intolerant can consume yoghurt without ill effects, because much of the lactose in the milk precursor is converted to lactic acid by the bacterial culture. Yoghurt containing live cultures is sometimes used in an attempt to prevent antibiotic-associated diarrhea. A study published in the *International Journal of Obesity* (11 January 2005) also found that the consumption of low-fat yoghurt can promote weight loss, especially due to the calcium in the yoghurt.

Varieties and Presentation

Dadiah, or Dadih, is a traditional West Sumatran yoghurt made from water buffalo milk. It is fermented in bamboo tubes.

Yoghurt is popular in Nepal, where it is served as both an appetiser and dessert. Locally called *dahi*, it is a part of the Nepali culture, used in local festivals, marriage ceremonies, parties, religious occasions, family gatherings, and so on. The most famous type of Nepalese yoghurt is called *juju dhau*, originating from the city of Bhaktapur. In Tibet, yak milk (technically dri milk, as the word yak refers to the male animal) is made into yoghurt (and butter and cheese) and consumed.

Tarator and Cacýk are popular cold soups made from yoghurt, popular during summertime in Albania, Bulgaria, Republic of Macedonia, and Turkey. They are made with ayran, cucumbers, dill, salt, olive oil, and optionally garlic and ground walnuts. Tzatziki, a thick yoghurt-based sauce similar in concoction to tarator, is popular in Greece. Bulgaria typically calls tzatziki "dry tarator".

Khyar w Laban (cucumber and yogurt salad) is a popular dish in Lebanon. Also, a wide variety of local Lebanese dishes are cooked with yogurt like "Kibbi bi Laban" etc.

Rahmjoghurt, a creamy yoghurt with much higher fat content (10%) than most yoghurts offered in English-speaking countries (*Rahm* is German for "cream"), is available in Germany and other countries.

Cream-top yoghurt is yoghurt made with unhomogenised milk. A layer of cream rises to the top, forming a rich yoghurt cream. Cream-top yoghurt was first made commercially popular in the United States by Brown Cow of Newfield, New York, bucking the trend toward low- and non-fat yoghurts.

Jameed is yoghurt which is salted and dried to preserve it. It is popular in Jordan.

Zabadi is the type of yoghurt made in Egypt, usually from the milk of the Egyptian water buffalo. It is particularly associated with Ramadan fasting, as it is thought to prevent thirst during all-day fasting.

Raita is a yoghurt-based South Asian/Indian condiment, used as a side dish. The yoghurt is seasoned with cilantro (coriander), cumin, mint, cayenne pepper, and other herbs and spices. Vegetables such as cucumber and onions are mixed in, and the mixture is served chilled. Raita has a cooling effect on the palate which makes it a good foil for spicy Indian dishes.

Dudh is a Sindhi-curd, popular in India. People drink dudh along with food at intervals, to help digestion and make food more delicious. In some places dudh is also served with plain rice.

Dahi is a yoghurt of the Indian subcontinent, known for its characteristic taste and consistency. The word *dahi* seems to be derived from the Sanskrit word *dadhi*, one of the five elixirs, or panchamrita, often used in Hindu ritual. *Dahi* also holds cultural symbolism in many homes in the *Mithilanchal* region of Bihar. It is found in different flavours, two of which are famous: sour yoghurt (*tauk doi*) and sweet yoghurt (*meesti* or *podi doi*). In India, it is often used in cosmetics mixed with turmeric and honey. Sour yoghurt is also used as a hair conditioner by women in many parts of India. *Dahi* is also known as *Thayiru* (Malayalam), *doi* (Assamese, Bengali), *dohi* (Oriya), *perugu* (Telugu), *Mosaru* (Kannada), *Thayir* (Tamil), or *Quzana a puuner* (Pashto).

Srikhand, a popular dessert in India, is made from drained yoghurt, saffron, cardamom, nutmeg and sugar and sometimes fruits such as mango or pineapple.

Sweetened and Flavoured Yoghurt

To offset its natural sourness, yoghurt can be sold sweetened, flavoured or in containers with fruit or fruit jam on the bottom. If the fruit has been stirred into the yoghurt before purchase, it is commonly referred to in the United States as Swiss-style. Most yoghurts in North America have added pectin, found naturally in fruit, and/or gelatin to artificially create thickness and creaminess at lower cost. This type of adulterated product is also marketed under the name Swiss-style, although it is unrelated to the way yoghurt is eaten in Switzerland. Some yoghurts, often called "cream line," are made with whole milk which has not been homogenised so the cream rises to the top. Fruit jam is used instead of raw fruit pieces in fruit yoghurts to allow storage for weeks.

Sweeteners such as cane sugar or sucralose – for low-calorie yogurts – are often present in large amounts in commercial yoghurt.

In the USA, sweetened, flavoured yoghurt is the most popular type, typically sold in single-serving plastic cups. Typical flavours are vanilla, honey, or fruit such as strawberry, blueberry, blackberry, raspberry, or peach. In recent years, some manufacturers are marketing flavours inspired by desserts, such as chocolate or cheesecake, with many variants.

In Australia, flavoured and Greek are the two most popular types of yoghurt, and are usually sold in one-litre tubs.

Strained Yoghurts

Strained yoghurts are types of yoghurt which are strained through a paper or cloth filter, traditionally made of muslin, to remove the whey, giving a much thicker consistency and a distinctive, slightly tangy taste.

Labneh is a strained yoghurt used for sandwiches popular in Arab countries. Olive oil, cucumber slices, olives, and various green herbs may be added. It can be thickened further and rolled into balls, preserved in olive oil, and fermented for a few more weeks. It is sometimes used with onions, meat, and nuts as a stuffing for a variety of pies or kebbeh balls.

Some types of strained yoghurts are boiled in open vats first, so that the liquid content is reduced. The popular East Indian dessert, a variation of traditional dahi called mishti dahi, offers a thicker, more custard-like consistency, and is usually sweeter than western yoghurts.

Strained yoghurt is also enjoyed in Greece and is the main component of *tzadziki* (from Turkish "cacik"), a well-known accompaniment to gyros and souvlaki pita sandwiches.

Beverages

Ayran or dhalla is a yoghurt-based, salty drink popular in Albania, Bulgaria, Turkey, Azerbaijan, Iran, Pakistan, the Republic of Macedonia, Kazakhstan and Kyrgyzstan. It is made by mixing yoghurt with water and (sometimes) salt. The same drink is known as *doogh* in Iran; *tan* in Armenia; *laban ayran* in Syria and Lebanon; *shenina* in Iraq and Jordan; *laban arbil* in Iraq; *majjiga* (Telugu), *majjige* (Kannada), and *moru* (Tamil and Malayalam) in South India; *lassi* in Punjab and all over India. A similar drink, doogh, is popular in the Middle East between Lebanon, Iran and Afghanistan; it differs from ayran by the addition of herbs, usually mint, and is sometimes carbonated, commonly with carbonated water. Lassi (urdu) is a yoghurt-based beverage originally from the Indian subcontinent that is usually slightly salty or sweet. Lassi is a staple of Punjab. In some parts of the subcontinent, the sweet version may be commercially flavoured with rosewater, mango or other fruit juice to create a very different drink. Salty lassi is usually flavoured with ground, roasted cumin and red chillies; this salty variation may also use buttermilk, and is interchangeably called *ghol* (Bengal), *mattha* (North India), *majjiga* (Andhra Pradesh), *moru* (Tamil Nadu and Kerala), *Dahi paani* (Odisha), *tak* (Maharashtra), or *chaas* (Gujarat). Lassi is also very widely drunk in Pakistan.

Sweetened yoghurt drinks are the usual form in Europe (including the UK) and the US, containing fruit and added sweeteners. These are typically called "drinking / drinkable yoghurt", such as Yop and BioBest Smoothie.

Also available are "yoghurt smoothies", which contain a higher proportion of fruit and are more like smoothies. In Ecuador, yoghurt smoothies flavoured with native fruit are served with pan de yuca as a common type of fast food.

Lassi

Lassi is a popular and traditional Punjabi yogurt-based drink of India and Pakistan. It is made by blending yogurt with water or milk and Indian spices. Traditional lassi (also known as salted lassi, or, simply lassi) is a savory drink sometimes flavoured with ground roasted cumin while sweet lassi on the other hand is blended with sugar or fruits instead of spices.

In Dharmic religions, yogurt sweetened with honey is used while performing religious rituals. Less common is lassi served with milk and is topped with a thin layer of clotted cream. Lassis are enjoyed chilled as a hot-weather refreshment, mostly taken with lunch. With a little turmeric powder mixed in, it is also used as a folk remedy for gastroenteritis.

Variations

Traditional Mild Salted Lassi

This form of lassi is more common in villages of Punjab & Porbandar, Gujarat (India). It is prepared by blending yogurt with water and adding salt and other spices to taste. The resulting beverage is known as salted lassi.

Sweet Lassi

Sweet lassi is a form of lassi flavoured with sugar, rosewater and/

or lemon, strawberry or other fruit juices. Saffron lassis, which are particularly rich, are a specialty of Sindh in Pakistan and Jodhpur and Rajasthan in India. *Makkhaniya lassi* is simply lassi with lumps of butter in it (*makkhan* is the Punjabi, Urdu, Hindi and Gujarati word for butter). It is usually creamy like a milkshake. Drinking sweet lassi can cause drowsiness, which might help people with sleep disorders.

Mango Lassi

Mango lassi is most commonly found in India and Pakistan though it is gaining popularity worldwide. It is made from yogurt, milk or water and mango pulp. It may be made with or without additional sugar. It is widely available in UK, Malaysia and Singapore, due to the sizable Pakistani/Indian minority, and in many other parts of the world. In various parts of Canada, mango lassi is a cold drink consisting of sweetened kesar mango pulp mixed with yogurt, cream, or ice cream. It is served in a tall glass with a straw, often with ground pistachio nuts sprinkled on top.

Bhang Lassi

Bhang (or bhung) lassi is a special lassi that contains *bhang*, a liquid derivative of cannabis (marijuana), which has effects similar to other eaten forms of marijuana. It is legal in many parts of India and mainly sold during Holi, when pakoras containing bhang are also sometimes eaten.

Rajasthan is known to have licensed bhang shops, and in many places one can buy bhang products and drink bhang lassis. However, the term "bhang lassie" is more often a misnomer, as bhang is almost always drunk with thandai, which does not include any curd (joghurt), but either milk, water, crushed ice, sugar, kesar pista flavouring (a ready made thandai syrup), saffron, ground almonds, spices, gound melon and poppy seeds and bhang...etc.

Chaas

Chaas or chaach is a salted drink like lassi; however, chaas contains more water than lassi and has the butterfat removed, so its consistency is not as thick as lassi. Salt and Jeera (cumin seeds) are normally added for taste and sometimes even fresh coriander. Fresh ground ginger & green chillies may also be added as seasoning.

Chaas is popular in the Pakistani/Indian Punjab regions of Bhakkar, Jalandhar, and D.I. Khan, and in the Indian states Gujarat and Rajasthan, where it is drunk with the main meal. It is known to aid digestion and is an excellent coolant in the Pakistani and Indian summers. It is called 'majjige' in Kannada, 'majjiga' in Telugu and 'moru' in Tamil and Malayalam.

Ayran

A drink in Turkey is similar to Lassi called Ayran. It is also made with yogurt and water. In Iran and Afghanistan a similar drink is called Doogh.

Tahn

Tahn is the Armenian version of the yogurt drink, consumed cold, with or without food. It is diluted with water and is flavoured with salt.

Cultural References

A 2008 print and television ad campaign for HSBC, written by Jeffree Benet of JWT Hong Kong, tells a tale of a Polish washing machine manufacturer's representative sent to India to discover why their sales are so high there. On arriving, the representative investigates a Lassi parlour, where he is warmly welcomed, and finds several washing machines being used to mix Lassi. The owner tells him he is able to "make ten times as much Lassi as I used to!"

On his *No Reservations* television program, celebrity chef Anthony Bourdain visited a "Govt Authorised" Bhang Shop in Jaisalmer Fort, Rajasthan. The proprietor offered him three varieties of bhang lassi: "normally strong, super duper sexy strong, and full power 24 hour, no toilet, no shower."

Buttermilk

Buttermilk refers to a number of dairy drinks. Originally, buttermilk was the liquid left behind after churning butter out of cream. It also refers to a range of fermented milk drinks, common in warm climates (e.g., Middle East, Pakistan, India, or the Southern United States) where fresh milk would otherwise sour quickly. It is also popular in Scandinavia and the Netherlands, despite the colder climates.

Buttermilk may also refer to a fermented dairy product produced from cow's milk with a characteristically sour taste caused by lactic acid bacteria. This variant is made in one of two ways:*cultured* buttermilk is made by adding lactic acid bacteria (*Streptococcus lactis*) to milk; *Bulgarian buttermilk* is created with a different strain of bacteria called *Lactobacillus bulgaricus*, which creates more tartness.

Whether traditional or cultured, the tartness of buttermilk is due to the presence of acid in the milk. The increased acidity is primarily due to lactic acid, a by-product naturally produced by lactic acid bacteria while fermenting lactose, the primary sugar found in milk. As lactic acid is produced by the bacteria, the pH of the milk decreases and casein, the primary protein in milk, precipitates causing the curdling or clabbering of milk. This process makes buttermilk thicker than plain milk. While both traditional and cultured buttermilk contain lactic acid, traditional buttermilk tends to be thinner whereas cultured buttermilk is much thicker.

Traditional Buttermilk

Originally, buttermilk was the liquid left over from churning butter from cream. Traditionally, before cream could be skimmed from whole milk, it was left to sit for a period of time to allow the cream and milk to separate. During this time, the milk would be fermented by the naturally occurring lactic acid-producing bacteria in the milk. This facilitates the butter churning process since fat from cream with a lower pH will coalesce more readily than that from fresh cream. The acidic environment also helps prevent potentially harmful microorganisms from growing, increasing shelf-life. However, in establishments that used cream separators, the cream would hardly be acid at all.

In the Indian subcontinent, buttermilk is taken to be the liquid leftover after extracting butter from churned yogurt (dahi). Today, this is called *traditional buttermilk*. Traditional buttermilk is still common in many Indo-Pakistani households but rarely found in western countries. In Southern India and most areas of the Punjab, buttermilk with added water, sugar and/or salt, asafoetida, and curry leaves is given at stalls in festival times.

Cultured Buttermilk

Commercially available cultured buttermilk is pasteurized and homogenized (if 1% or 2% fat) milk which has been inoculated with a culture of lactic acid bacteria to simulate the naturally occurring bacteria found in the old-fashioned product. Some dairies add colored flecks of butter to cultured buttermilk to simulate the residual pieces of butter that can be left over from the churning process of traditional buttermilk.

Buttermilk solids have increased in importance in the food industry. Such solids are used in ice cream manufacture. Adding specific strains of bacteria to pasteurized milk allows more consistent production.

In the early 1900s, cultured buttermilk was labelled *artificial buttermilk*, to differentiate it from traditional buttermilk, which was known as *natural* or *ordinary buttermilk*.

Acidified buttermilk is a related product that is made by adding a food-grade acid (such as lemon juice) to milk.

Probiotics

Probiotics are live microorganisms thought to be beneficial to the host organism. According to the currently adopted definition by FAO/WHO, probiotics are: "Live microorganisms which when administered in adequate amounts confer a health benefit on the host". Lactic acid bacteria (LAB) and bifidobacteria are the most common types of microbes used as probiotics; but certain yeasts and bacilli may also be helpful. Probiotics are commonly consumed as part of fermented foods with specially added active live cultures; such as in yogurt, soy yogurt, or as dietary supplements.

Etymologically, the term appears to be a composite of the Latin preposition *pro* ("for") and the Greek adjective biotic, the latter deriving from the noun *âβïò* (bios, "life"). At the start of the 20th century, probiotics were thought to beneficially affect the host by improving its intestinal microbial balance, thus inhibiting pathogens and toxin producing bacteria. Today, specific health effects are being investigated and documented including alleviation of chronic intestinal inflammatory diseases, prevention and treatment of pathogen-induced diarrhea, urogenital infections, and atopic diseases.

History

The original observation of the positive role played by certain bacteria was first introduced by Russian scientist and Nobel laureate Eli Metchnikoff, who in the beginning of the 20th century suggested that it would be possible to modify the gut flora and to replace harmful microbes with useful microbes. Metchnikoff, at that time a professor at the Pasteur Institute in Paris, produced the notion that the aging process results from the activity of putrefactive (proteolytic) microbes producing toxic substances in the large bowel. Proteolytic bacteria such as clostridia, which are part of the normal gut flora, produce toxic substances including phenols, indols and ammonia from the digestion of proteins. According to Metchnikoff these compounds were responsible for what he called "intestinal auto-intoxication", which caused the physical changes associated with old age.

It was at that time known that milk fermented with lactic-acid bacteria inhibits the growth of proteolytic bacteria because of the low pH produced by the fermentation of lactose. Metchnikoff had also observed that certain rural populations in Europe, for example in Bulgaria and the Russian steppes who lived largely on milk fermented by lactic-acid bacteria were exceptionally long lived. Based on these facts, Metchnikoff proposed that consumption of fermented milk would "seed" the intestine with harmless lactic-acid bacteria and decrease the intestinal pH and that this would suppress the growth of proteolytic bacteria. Metchnikoff himself introduced in his diet sour milk fermented with the bacteria he called "Bulgarian Bacillus" and found his health benefited. Friends in Paris soon followed his example and physicians began prescribing the sour milk diet for their patients.

Bifidobacteria were first isolated from a breast-fed infant by Henry Tissier who also worked at the Pasteur Institute. The isolated bacterium named Bacillus bifidus communis was later renamed to the genus *Bifidobacterium*. Tissier found that bifidobacteria are dominant in the gut flora of breast-fed babies and he observed clinical benefits from treating diarrhea in infants with bifidobacteria. The claimed effect was bifidobacterial displacement of proteolytic bacteria causing the disease.

During an outbreak of shigellosis in 1917, German professor Alfred Nissle isolated a strain of *Escherichia coli* from the feces of a soldier who was not affected by the disease. Methods of treating infectious diseases were needed at that time when antibiotics were not yet available, and Nissle used the *Escherichia coli* Nissle 1917 strain in acute gastrointestinal infectious salmonellosis and shigellosis.

In 1920, Rettger demonstrated that Metchnikoff's "Bulgarian Bacillus", later called *Lactobacillus delbrueckii subsp. bulgaricus*, could not live in the human intestine, and the fermented food phenomena petered out. Metchnikoff's theory was disputable (at this stage), and people doubted his theory of longevity.

After Metchnikoff's death in 1916, the centre of activity moved to the United States. It was reasoned that bacteria originating from the gut were more likely to produce the desired effect in the gut, and in 1935 certain strains of *Lactobacillus acidophilus* were found to be very active when implanted in the human digestive tract. Trials were carried out using this organism, and encouraging results were obtained especially in the relief of chronic constipation.

The term "probiotics" was first introduced in 1953 by Kollath. Contrasting antibiotics, probiotics were defined as microbially derived factors that stimulate the growth of other microorganisms. In 1989 Roy Fuller suggested a definition of probiotics which has been widely used: "*A live microbial feed supplement which beneficially affects the host animal by improving its intestinal microbial balance*". Fuller's definition emphasizes the requirement of viability for probiotics and introduces the aspect of a beneficial effect on the host.

In the following decades intestinal lactic acid bacterial species with alleged health beneficial properties have been introduced as probiotics, including *Lactobacillus rhamnosus*, *Lactobacillus casei*, and *Lactobacillus johnsonii*.

Benefits

Experiments into the benefits of probiotic therapies suggest a range of potentially beneficial medicinal uses for probiotics. For many of the potential benefits, research is limited and only preliminary results are available. It should be noted that the effects described are *not* general effects of probiotics. Recent research on the molecular biology and genomics of *Lactobacillus* has focused on the interaction with the immune system, anti-cancer potential, and potential as a biotherapeutic agent in cases of antibiotic-associated diarrhoea, travellers' diarrhoea, pediatric diarrhoea, inflammatory bowel disease and irritable bowel syndrome.

All effects can only be attributed to the individual strain(s) tested. Testing of a supplement does not indicate benefit from any other strain of the same species, and testing does not indicate benefit from the whole group of LAB (or other probiotics).

Managing Lactose Intolerance

As lactic acid bacteria actively convert lactose into lactic acid, ingestion of certain active strains may help lactose intolerant individuals tolerate more lactose than they would have otherwise.

Preventing Colon Cancer

In laboratory investigations, some strains of LAB (*Lactobacillus bulgaricus*) have demonstrated anti-mutagenic effects thought to be due to their ability to bind with heterocyclic amines, which are carcinogenic substances formed in cooked meat. Animal studies have demonstrated that some LAB can protect against colon cancer in rodents, though human data is limited and conflicting. Most human trials have found that the strains tested may exert anti-carcinogenic effects by decreasing the activity of an enzyme called â-glucuronidase (which can generate carcinogens in the digestive system). Lower rates of colon cancer among higher consumers of fermented dairy products have been observed in one population study.

Lowering Cholesterol

Animal studies have demonstrated the efficacy of a range of LAB to be able to lower serum cholesterol levels, presumably by breaking down bile in the gut, thus inhibiting its reabsorption (which enters the blood as cholesterol). Some, but not all human trials have shown that dairy foods fermented with specific LAB can produce modest reductions in total and LDL cholesterol levels in those with normal levels to begin with, however trials in hyperlipidemic subjects are needed.

Lowering Blood Pressure

Several small clinical trials have indicated that consumption of milk fermented with various strains of LAB may result in modest reductions in blood pressure. It is thought that this is due to the ACE inhibitor-like peptides produced during fermentation.

Improving Immune Function and Preventing Infections

LAB are thought to have several presumably beneficial effects on immune function. They may protect against pathogens by means of competitive inhibition (i.e., by competing for growth) and there is evidence to suggest that they may improve immune function by increasing the number of IgA-producing plasma cells, increasing or improving phagocytosis as well as increasing the proportion of T lymphocytes and Natural Killer cells.

Clinical trials have demonstrated that probiotics may decrease the incidence of respiratory tract infections and dental caries in children. LAB foods and supplements have been shown to aid in the treatment and prevention of acute diarrhea, and in decreasing the severity and duration of rotavirus infections in children and travelers' diarrhea in adults.

A 2010 study suggested that the anecdotal benefits of probiotic therapies as beneficial for preventing secondary infections, a common complication of antibiotic therapy, may be because keeping the immune system primed by eating foods enhanced with "good" bacteria may help counteract the negative effects of sickness and antibiotics. It was thought that antibiotics may turn the immune system "off" while probiotics turns it back on "idle", and more able to quickly react to new infections.

Helicobacter Pylori

LAB are also thought to aid in the treatment of *Helicobacter pylori* infections (which cause peptic ulcers) in adults when used in combination with standard medical treatments. However more studies are required into this area.

Antibiotic-associated Diarrhea

Antibiotic-associated diarrhea (AAD) results from an imbalance in the colonic microbiota caused by antibiotic therapy. Microbiota alteration changes carbohydrate metabolism with decreased short-chain fatty acid absorption and an osmotic diarrhea as a result. Another consequence of antibiotic therapy leading to diarrhea is overgrowth of potentially pathogenic organisms such as *Clostridium difficile*.

Probiotic treatment can reduce the incidence and severity of AAD as indicated in several meta-analyses. However, further documentation of these findings through randomized, double blind, placebo-controlled trials are warranted. Efficacy of probiotic AAD prevention is dependent on the probiotic strain(s) used and on the dosage. Up to a 50% reduction of AAD occurrence has been found. No side-effects have been reported in any of these studies. Caution should, however, be exercised when administering probiotic supplements to immunocompromised individuals or patients who have a compromised intestinal barrier.

Reducing Inflammation

LAB and supplements have been found to modulate inflammatory and hypersensitivity responses, an observation thought to be at least in part due to the regulation of cytokine function. Clinical studies suggest that they can prevent reoccurrences of inflammatory bowel disease in adults, as well as improve milk allergies.

They are not effective for treating eczema, a persistent skin inflammation. How probiotics counteract immune system overactivity remains unclear, but a potential mechanism is desensitization of T lymphocytes, an important component of the immune system, towards pro-inflammatory stimuli .

Improving Mineral Absorption

It is hypothesized that probiotic lactobacilli may help correct malabsorption of trace minerals, found particularly in those with diets high in phytate content from whole grains, nuts, and legumes.

Preventing Harmful Bacterial growth Under Stress

In a study done to see the effects of stress on intestinal flora, rats that were fed probiotics had little occurrence of harmful bacteria latched onto their intestines compared to rats that were fed sterile water.

Treating Irritable Bowel Syndrome and Colitis

B. infantis 35624, sold as Align, was found to improve some symptoms of irritable bowel syndrome in women in a recent study. Another probiotic bacterium, *Lactobacillus plantarum* 299v, was also found to be effective in reducing IBS symptoms. Additionally, a probiotic formulation, VSL#3, was found to be safe in treating ulcerative colitis, though efficacy in the study was uncertain. *Bifidobacterium animalis* DN-173 010 may help. For maintenance of remission of ulcerative colitis, Mutaflor (*E.coli* Nissle 1917) there are 3 controlled, randomized, double blind clinical studies which have proven equivalence of Mutaflor and mesalazine (5-ASAs).

Managing Urogenital Health

Several in vitro studies have revealed probiotics' potential in relieving urinary tract infections and bacterial vaginosis. Results have been varied on these studies, and in vivo studies are still required in this area to determine efficacy.

Other

A study in 2004 testing the immune system of students given either milk or Actimel over a 6 week exam period (3 weeks of studying, 3 weeks of exams) tested 19 different biomarkers. Of these 19 biomarkers only 2 were shown to be different between the two groups, increased production of lymphocytes and increased production of CD56 cells. The tests were not blind and show that certain probiotic strains may have no overall effect on the immune system or on its ability.

A 2007 study at University College Cork in Ireland showed that a diet including milk fermented with *Lactobacillus* bacteria prevented *Salmonella* infection in pigs. A 2007 clinical study at Imperial College London showed that preventive consumption of a commercially available probiotic drink containing *L casei DN-114001, L bulgaricus,* and *S thermophilus* can reduce the incidence of antibiotic-associated diarrhea and *C difficile*-associated diarrhea. The efficacy and safety of a daily dose of *Lactobacillus acidophilus* CL1285 in the prevention of AAD was demonstrated by Montreal's Maisonneuve-Rosemont Hospital, in a clinical study of hospitalized patients.

Current research is focusing on the molecular biology and genomics of *Lactobacillus* and bifidobacteria. The application of modern whole genome approaches is providing insights into bifidobacterial evolution, while also revealing genetic functions that explain their presence in the particular ecological environment of the gastrointestinal tract.

Probiotics are used in industry to improve yields of pork and chicken production.

Disadvantages

In some situations, such as where the person consuming probiotics is critically ill, probiotics could be harmful. In a therapeutic clinical trial conducted by the Dutch Pancreatitis Study Group, the consumption of a mixture of six probiotic bacteria, increased the death rate of patients with predicted severe acute pancreatitis.

In a clinical trial conducted at the University of Western Australia, aimed at showing the effectiveness of probiotics in reducing childhood allergies, Dr Susan Prescott and her colleagues gave 178 children either a probiotic or a placebo for the first six months of their life. Those given the good bacteria were more likely to develop a sensitivity to allergens.

Some hospitals have reported treating lactobacillus septicaemia, which is a potentially fatal disease caused by the consumption of probiotics by people with lowered immune systems or who are already very ill. There is no published evidence that probiotic supplements are able to replace the body's natural flora when these have been killed off; indeed bacterial levels in feces disappear within days when supplementation ceases.

Probiotics taken orally can be destroyed by the acidic conditions of the stomach. A number of micro-encapsulation techniques are being developed to address this problem. Recent studies indicate that probiotic products such as yogurts could be a cause for obesity trends. However, this is contested as the link to obesity and other health related issue with yogurt may link to its dairy attributes.

Some experts are skeptical on the efficacy of many strains and believe not all subjects will benefit from the use of probiotics. A criticism of probiotic supplements is the cost and value of probiotics products.

Strains

Live probiotic cultures are available in fermented dairy products and probiotic fortified foods. However, tablets, capsules, powders and sachets containing the bacteria in freeze dried form are also available.

Some additional forms of yogurt bacteria include:

- *Lactobacillus bulgaricus*
- *Streptococcus thermophilus*
- *Lactobacillus bifidus* - became new genus *Bifidobacterium*.

Some fermented products containing similar lactic acid bacteria include:

- Pickled vegetables
- Fermented bean paste such as tempeh, miso and doenjang
- Kefir
- Buttermilk or Karnemelk
- Kimchi
- Pao cai
- Sauerkraut
- Soy sauce
- Zha cai.

Multi-probiotic

Research is emerging on the potential health benefits of multiple probiotic strains as a health supplement as opposed to a single strain. The human gut is home to some 400-500 types of microbes. It is thought that this diverse environment may benefit from multiple probiotic strains; different strains populate different areas of the digestive tract, and studies are beginning to link different probiotic strains to specific health benefits.

Chapter 2

Physico-chemical Testing of Milk and Dairy Products

Microbiological and physicochemical analysis of different UHT milks available in market Raw milk is milk in its natural (unpasteurized) state. Contaminated raw milk can be a source of harmful bacteria, such as those that cause undulant fever, dysentery, salmonellosis and tuberculosis. "Certified" milk, obtained from cows certified as healthy, is unpasteurized milk with a bacteria count below a specified standard, but it still can contain significant numbers of disease producing organisms.

Different heat and treatments are given to raw milk in order to remove pathogenic organisms, to increase the shelf life, to help subsequent processing e.g. for warming before separation and homogenization or as an essential treatment before cheese making, yoghurt manufacture and production of evaporated and dried milk products (Singh, 1993).

Pasteurization, sterilization (in bottle) and UHT (ultra-high-temperature) treatment integrated with aseptic packing. Sterilization (in bottle) is the term applied to a heat treatment process which has a bactericidal effect greater than pasteurization. Although it does not result in sterility, it gives the processed milk a longer shelf life. As a result of the long holding time at this elevated temperature, the product has a cooked flavour and a pronounced brown colour.

Unlike sterilization, pasteurization is not intended to kill all pathogenic micro-organisms in the food or liquid. Instead, pasteurization aims to reduce the number of viable pathogens so they are unlikely to cause disease. Ultra-high temperature (UHT or ultra-heat treated) is also used for milk treatment. UHT processing holds the milk at a temperature of 138°C (250°F) for a fraction of a second. Milk simply labelled "pasteurization" is usually treated with the HTST method, whereas milk labelled "ultra-pasteurization" or simply "UHT" has been treated with the UHT method (Bylund, 1995). Heating of milk accounts 2 main problems, age gelation and off flavour development, which limits shelf life of milk. UHT treatment of milk leads to a much larger production of small sized casein micelles compared to raw or pasteurized milk (Singh, 1993).

Biochemical processes involve are heat resistance and reactivation of natural and bacterial proteases and survival of bacterial spores. Proteolysis of UHT milk during storage at room temperature is a major factor limiting the shelf life through changes in its flavour and texture (Datta et al., 2002).

The changes ultimately reduce the quality and limit the shelf life of UHT milk via development of off flavours, fat separation and sedimentation, which principally falls into 2 categories, liberation of volatile fatty acids such as butyric acid and oxidation of free or unsaturated fatty acids (Datta et al., 2002). Above 135°C the protein deposited on the fat globule membrane form a network which makes the membrane denser and less permeable.

There is an increase in acidity and viscosity with a decrease in pH with the storage time increased both in UHT. Clare et al. (2005) determined that sweet aromatic flavour and sweet taste of UHT milk decreases during storage.

The microorganisms, which cause spoilage in milk, which is intended to be sterile (UHT treatment), are either resistant types that have survived the heat treatment, or organisms that have contaminated the product after the sterilization process. Contamination may either be by heat labile organism or heat resistant forms such as spores. Contaminating spores are, however, likely to be less heat resistant than those, which might survive the heat treatment.

The problem of post treatment contamination of in container sterilized product is well known. The contamination can either through poor seal or through pinhole in the container. Post treatment contaminants in UHT milk may be either spores, which would not be expected to be heat resistant enough to survive the heat treatment or non heat resistant vegetative organisms. Organisms of first type will probably have entered from ineffectively sterilized plant down stream from the heat treatment stage of the process, which includes spores of *Bacillus cereus* and *Bacillus licheniformis*. Organisms of second type will probably have entered through poorly sealed container after aseptic filling.

The types of spores, which have been investigated as of particular relevance in the UHT, are those of *Bacillus stearothermophilus, Bacillus subtilis* and *Clostridium botulinum* has been studied.

The high spore counts can occur at the dairy farm and that feed and milking equipment can act as reservoirs or entry points for potentially highly heat resistant spores into raw milk. Lowering this spore load by good hygienic measures could probably further reduce the contamination level of raw milk, in this way minimizing the aerobic spore forming bacteria that could lead to spoilage of milk and dairy products (Westhoff and Dougerty, 1981).

These problems had been reported internationally since long, hence the project was planned to observe the physicochemical. In this study the de-clared shelf life of different UHT milk available in market is studied.

Materials and Methods

The samples were taken in sterilized syringes for microbiological analysis and in clean stainless steel containers of 1 liter for chemical and sensory analysis. The samples were analyzed at interval of 1, 2, 3, 4, 5, 6, 7, 8, 9, 10, 11 and 12 weeks. During this period, samples were stored at room temperature (25°C) to provide them similar conditions, as they are stored in market. For microbiological analysis the samples were examined for aerobic plate counts (APC), *E. coli* counts, *B. cereus, B. subtilis* counts and for spore formers counts. The parameters examined for the chemical analysis were sedimentation, pH, and acidity as lactic acid %, fat % before and after shaking the milk, SNF % before and after shaking and protein % before and after shaking. For sensory evaluation colour, aroma and taste were examined.

Physicochemical Analysis of Milk

To assess the physical and chemical changes in processed milk samples following tests were carried out.

Sedimentation Test

Sedimentation test was performed by following modified method as described by Ramsey and Swartzel (1984). According to this method, milk was drain from the cartons leaving the bottom 4 cm. The cartons were inverted for approximately 10 min, up righted and placed in the exhaust hood to dry. The cartons were allowed to dry for 48 h after the bottom flaps or wings of cartons had been opened to facilitate the drying of any sediment entrapped there. The cartons were weighted and then washed thoroughly to remove any sediment or residue adhering to the container. The washed cartons were again dried and weighted.

Solids Non Fats (SNF) %

Solids non fats (SNF) % was determined by lactometeric method as described by Ramsey and Swartzel (1984).

Total Titratable Acidity

Total titratable acidity determined according to the method of AOAC, (2005).

pH

The pH value of milk was determined by using a digital pH meter (AOAC, 2005). Prior to use, the pH meter was standardized with standard buffer solution of pH 4 and 7.

Fat

Milk fat % was determined by Gerber (1997) method as described by FAO (1997) by using the butyrometer.

Protein

The protein was estimated by formal titration method (Davide, 1977).

Microbial Analysis

Microbial analysis was performed according to standard methods (AOAC, 2005).

Total Viable Counts

The plate count agar media (Bridson, 1995) was used for the total viable count in UHT milk samples (AOAC, 2005). Plates were incubated for 24 h at 37°C.

Determination of Coliforms

Coliform counts were determined by pour plate method on violet red bile agar, prepared according to the manufacturer instructions. All plates were incubated at 37°C for 24 h.

Determination of Bacillus Species

B. cereus selective agar base (Bridson, 1995) is used for isolation and enumeration of *B. cereus* and *B. subtilis*. All plates were incubated at 37°C for 24 h.

Determination of Spore Formers

Plate count agar media (Bridson, 1995) is used for the enumeration of spore formers. Sterile medium was poured into sterile petri plates and allowed the medium to solidify. Sample is heated at 80°C for 10 min using water bath. These plates were inoculated with 1 ml sample by using sterile pipette. After inoculation, the sample was well mixed in the petri plates by to and fro motion. All plates were incubated in an inverted position for 72 h at 55°C.

E. coli Counts

For *E. coli* count MacConkey's agar (Bridson, 1995) was used. Sample from lactose positive tubes in case of coliform counts were applied directly on the MacConkey's agar (Bridson, 1995) plates and incubated at 37°C for 24 h.

Sensory analysis

The stored milk samples were evaluated sensorial for colour and flavor by scoring method as described by Larmond (1977).

Results and Discussion

The changes that have taken place during storage depend on temperature of storage, extent of exposure of the milk to light and availability of oxygen.

The dairy company of Pakistan shows shelf life of 12 weeks on the labels of milk packs, during this mentioned period. Milk must be in best condition for consumption. For storage time than a week or 2, these effects may be greater than those of the heat treatment.

Changes in colour, flavour and texture are readily detected by the consumer and may reduce the acceptability of the products. Other changes cannot be recognized by the consumer and are not necessarily correlated with organoleptic, recognizable changes, but are of potential nutritional importance. The quality of sediment depends on the raw milk and on the type and severity of the heat treatments. For any 1 type of process, the amount of sediments increases in the severity of the heat treatment. The amount of sediment decreases with homogenization pressure (Robinson, 1994). Results obtained from sedimentation test in UHT milk during storage period of 3 months (12 weeks) shows that there is an effect of heat processing and subsequent storage on sedimentation in all 4 samples of UHT milk. The changes started in week 2 of shelf life for samples I and III and sample II showed formation of sediments after week 6.

Sample III reaches up to 7.10/250 gml-1, which is a considerable changes and sample II showed formation of sediments after week 5. The alcohol test can be used to detect raw milk that is likely to give high level of the normal type of sediments and there are indications that it may be useful in predicting the abnormal type (Sweetsur and White, 1975). Processing operations influences acid base equilibrium in milk. UHT treatment results in a pH decrease, due to conversion of lactose into different organic acids. In milk, casein micelles are stable at natural pH, that is, 6.7.

Lowering the pH facilitates aggregations of casein micelles and forms a gel. Results regarding effect of storage on pH of UHT processed milk during storage period of 90 days show that there is storage effect on pH level. Maximum pH value (6.81and 6.85 in samples 1 and 2, respectively and 6.75 in samples III and IV) was recorded in 1st week while minimum pH values obtained in 12th week of shelf life (6.20, 6.65, 6.17 and 6.55 in sample 1, 2, 3 and 4 of UHT milk respectively). Vankatachalm and McMahon (1991) verified drop in pH and they associated it with browning reactions.

Andrews et al. (1977) confirmed similar effects and concluded that the level and extent of pH decrease was related to age gelation. When milk is heated at a temperature above 100oC and subsequent stored, lactose is degraded to acids. Formic acid is the principal acid produced due to which titratable acidity of milk rises.

Increase in free fatty acids is also responsible for increasing the total titratable acidity of milk (Swartzel, 1983). Results obtained by the analysis for total titratable acidity. The acidity value was 0.11% while during storage of UHT milk minimum acidity was recorded in 1st week and maximum value (0.18%) at 90 days life in sample 1 while 0.15 in case of sample 1 and 0.13 in samples 2 and 4. The proteins of milk are the constituents most affected by heating and subsequent storage of milk.

The principal changes in UHT milk during storage may be due to enzymes. Many proteins in milk are very heat labile e.g. whey proteins, vitamin binding protein, antimicrobial proteins etc. These proteins coagulate after heating hence the texture of milk is deteriorated during storage (Fox and McSweeny, 1998).

Casein polymerization is greater at high storage temperature, but occurs significantly even under refrigeration: 50% of the protein may be in the polymer form after 6 months at 37°C, and 21% after 6 months at 4°C (Andrews, 1977). The results regarding protein %

of stored UHT milk describes that there is effect of storage on protein contents of UHT processed milk. That is, in week 1 protein contents were 3.30%, 3.70% for sample 1 and 2 while in week 12 of storage were 2.35 and 3.48 respectively. In case of samples III and IV, protein contents were 3.40 in week 1 while it changes to 1.15 and 2.59, respectively, in week 12.

There is no change in protein % in all samples after shaking of UHT milk. Chen et al. (2005) showed almost a 90% loss and denaturation of S-lactoglobulin (LG) of the UHT processed and dry milks by using polyacrylamide gel electrophoresis. Of the principle constituent, the fats are probably least affected by UHT treatment.

It is concluded from the whole study that there is an increase in sedimentation value, fat separation, titratable acidity during storage, while decrease was found in pH and protein % during storage of 12 weeks. The increase in sedimentation shows excessive protein denaturation during processing and subsequent storage. In UHT processed milks the fat separation was observed during storage.

This high % of fat separation is attributed with less homogenizing efficiency during processing. On microbiological examination, not any colony found on TPC plates, *coliform* agar plates, *E. coli* plate, *B. cereus*, *B. subtilus*, and spore formers plates, in all the 4 samples of UHT milk during storage of 12 weeks.

Sensory characteristics showed a significant decrease in scores during storage. These all are factors that limit the shelf life of UHT milk. The shelf life of milk mainly depend on the quality of raw milk and better quality of milk can be achieved in Pakistan, when the manufacturers have better milk collection system.

The manufacturer of sample II has its own sophisticated type of milk collection system said to be VMCs (village milk collection centers). At these centers, milk is collected at small scale and in short time it is transported at low temperature to the processing plant, avoiding contamination, due to this practice the microbial as well as other contamination can be controlled in better way before heat treatment or processing.

While manufacturers of other dairy industries of Pakistan get milk from contractors and ice added milk is mostly supplied to these industries which disturb the mineral balance and natural emulsion and give higher water activity which leads to physicochemical, microbiological as well as sensory changes during shelf life of milk.

Microbiological and Physicochemical Properties of Raw Milk Used for Processing Pasteurized Milk in Blue Nile Dairy Company (Sudan)

Milk is a highly nutritious food, ideal for microbial growth and the fresh milk easily deteriorates to become unsuitable for processing and human consumption (FAO 2001). High bacterial counts are indicator of poor production hygiene or ineffective pasteurization of milk (Harding 1999).

Milk and milk products derived from dairy cows milk can harbour a variety of microorganisms and can be important sources of foodborne pathogens. The presence of food-borne pathogens in milk is due to direct contact with contaminated sources in the dairy farm environment and to excretion from the udder of an infected animal (El Zubeir *et al.* 2006).

The hygienic quality problems of milk may arise from raw milk of diseased animals (Murphy and Boor 2000). Kang *et al.* (2005) reported that the presence of antimicrobial substances in raw milk could have serious toxicological and technical consequences. Raw milk may contain over 2,000,000 cfu/ml before processing of liquid milk or cheese making (Kameni *et al.* 2002). The raw milk distributed for consumption in Sudan does not find the real quality control measures needed to be of good quality food (Mohamed and El Zubeir 2007). However, some new private dairy plants started the processing of fluid milk and some dairy products. These are faced with many problems of which the quality control measures constitute an important concern. Hence, the present study was designed to assess the chemical, physical and microbial properties of raw milk supplied to the Blue Nile Dairy Company plant (CAPO) and to compare it with the produced pasteurized milk.

Materials and Methods

Source of Milk: This study was carried out during June to September 2005, and the raw milk for this study was collected from two dairies, namely Blue Nile Dairy Company and Kordi Farms. Blue Nile Dairy Company plant deals with both suppliers as a source of raw milk for processing pasteurized milk.

Raw milk from both dairy farms is usually mixed before processing. Milk fat was standardized to 3-3.2%, and the milk was pasteurized at 72- 76< C for 15 second using a high temperature short time (HTST) plate heat exchanger (Wincantor Pasteurizer, Wincanton Engineering

Ltd, South Street Sherborne Dorset, U.K.). The pasteurized milk was packed into Tetra Pack container (Tetra Pack Technical Services AB, Ruben Rausing gata, 5-221 86 Lund, Sweden).

Raw milk samples (36 samples) and pasteurized milk after processing and before packaging (12 samples) were examined for total bacterial counts (TBC), coliform, psychrotrophic (PC) and thermoduric bacterial counts. Physiochemical properties (fat, protein, lactose, ash, solid not fat, density, acidity, pH and freezing point), antibiotics and phosphatase test were also done.

Chemical Analysis of Milk Samples: The milk constituents (fat, protein, lactose, ash and solid not fat) and physical characteristics (density and freezing point), of the milk samples were determined by milk analyzer Lactoscan 90 (Aple Industries services–La Roche Sur Foron, France), according to manufacturer's instructions.

Milk samples were mixed gently 4-5 times to avoid any air enclosure in the milk. Then 25 ml samples were taken in the sample-tube and put in the sample- holder one at a time with the analyzer in the recess position. Then when the starting button activated, the analyzer sucks the milk, makes the measurements, returns the milk in the sample-tube and the digital indicator (IED display) shows the specified results.

Antibiotic residues were determined using Delvotest® SP- ampule Kit (202– Delvotest SP 100, test box, DSM Food Specialties, the Netherlands). The method was carried out according to the manufacturer's instructions. Phosphatase test was done using Lactognost tables and powders obtained from Heyl, Chem. Pharm-Fabrik, 14167 Berlin.

The procedure for phosphatase test was done according to the manufacturer's instructions. The acidity of the samples was determined according to AOAC (1990). The temperature and the pH of the samples were determined using pH– meter (Wagtech, HI 8314 membrane pH Meter, U.K.).

Microbiological Analysis: The samples were examined for TBC, coliforms, thermoduric and psychrotrophic counts according to Houghtby et al. (1992); Christen et al. (1992); Ballou et al. (1995); Ravanis and Lewis (1995), respectively. Plate count agar No. 298 (Biomark Laboratories) was used for enumeration of TBC, thermoduric bacteria and psychrotrophic counts, while violet red bile agar No. 779 (Biomark Laboratories) was used to determine coliform counts.

The media were prepared according to manufacturer's instructions. Plates for enumeration of TBC, thermoduric bacteria and coliforms were incubated at 32° C for 48 hours, 37° C for 48 hours and 37° C for 24 hours, respectively. Plates for enumeration of psychrotrophic counts were incubated at 7° C for ten days. Developed colonies were counted using manual colony counter. The plates counting 25-250 colonies were selected as described by Houghtby *et al.* (1992). The number reciprocal of the dilution factor was recorded as colony forming unit per ml (cfu/ ml). The milk ring test (MRT) for brucellosis was carried out according to Harrigan and McCance (1976).

Statistical Analysis: Data were analyzed by SPSS programme (Statistical Package for Social Science, version 10.00). This test combines ANOVA with comparison of differences between means of the treatments at the significance level of P< 0.05.

Results and Discussion

The means of fat, protein, lactose, ash and solids not fat content were 4.14%, 3.48 %, 4.33%, 0.778% and 8.58% in raw milk samples mixture. The density, freezing point, titratable acidty and pH revealed 1.031, -0.520, 0.145 and 7.02. The analysis of variance showed highly significant variations (P< 0.01) due to the source of raw milk samples, except for fat.

The composition of raw milks in the present study was compared favourably with the composition of milk in northern Europe, which contained fat of 4.3%, total protein of 3.4%, lactose of 4.65%, ash of 0.73%, TS of 13.3% and SNF of 9.0% (Invensys APV 2002). This result also agrees with that reported by El Zubeir *et al.* (2005) for raw milk. The present study revealed lower mean values for lactose (%) than that reported by El Zubeir *et al.* (2005). The lower lactose may be due to the effect of psychotrophic bacteria (Ballou *et al.* 1995). The results of physicochemical analysis of mixed raw milk used for producing fluid milk and the pasteurized milk.

These results were higher compared to that reported by Elmagli and El Zubeir (2006a). These differences in milk composition may be due to initial raw milk used and the procedure of processing.

However the results of freezing point agreed with those reported by Tetra Pak Processing Systems (2003) for freezing points of raw and pasteurized milk -0.520± 0.001 and -0.447± 0.000, respectively obtained during the present study. This study also agreed with that reported by Elmagli and El Zubeir (2006a) who found the freezing point was

-0.4734± 0.05032 C. The obtained data for acidity o of raw and pasteurized milk of 0.145% and 0.143%, respectively, which are in line with that reported by Harding (1999), while the mean value was lower than that reported by Mohamed and El Zubeir (2007).

The microbiological quality of the raw milk used for processing pasteurized milk showed that the initial quality was good for TBC (log 4.800 cfu\ml), coliform counts (log 4.157 cfu/ml), thermoduric bacterial counts (log 2.994) and psychotrophic bacterial counts (log 810 cfu/ml). The analysis of variance showed highly significant differences (P< 0.01) due to the source of raw milk samples for TBC and coliform counts. This result was lower than that reported by PMO (2001) for the average standard plate counts for can and bulk milk (700.000 bacteria /ml and 100.0 bacteria /ml, respectively). Moreover, the microbial standards for grade A raw milk is 100.0 bacteria/ml (PMO, 2001).

The lower counts of bacteria may be due to good cleaning system and good handling from farms to the plant as was stated before by Chye *et al.* (2004). Lower TBC value was obtained for pasteurized milk than that reported by Elmagli and El Zubeir (2006b), who reported a range of 6.5×105 to 6.5×1014, but was similar to that of Reena *et al.* (2003). In addition, PMO (2001) reported that the bacteria standards for grade A pasteurized milk should be less than 20,000 bacteria /ml. Coliform bacteria counts of pasteurized milk showed lower numbers than these reported by Elmagli and El Zubeir (2006b).

The lower coliform counts might be due to hygienic quality of raw milk, proper pasteurization process, good packaging and good storage conditions. This agreed with PMO (2001) who reported that the total bacterial standards for grade A pasteurized milk should be < 10 coliform/ ml. In addition, coliform counts obtained are in line with Sudanese Standards (SSMO, 2005) which stated that the maximum coliform counts should not to exceed 102 cfu/ml.

Thermoduric bacterial counts (log 0.621cfu/ml) was lower than that reported by Mohamed and El Zubeir (2007). However, the present findings agreed with that reported by Invensys APV (2002) who reported an aerobic spores-forming bacteria of <400. The mean value of psychrotrophic bacteria for pasteurized milk was log 0.360 cfu/ml, which was lower counts compared with that reported by Elmagli and El Zubeir (2006b), who found the psychotrophic bacterial counts were <6.5×10 for pasteurized milk. All the samples during storage showed the absence of the phosphatase test.

This result might be due to proper pasteurization. However, Elmagli and El Zubeir (2006a) demonstrated that 10 % of the pasteurized milk samples were positive to the phosphatase test. No brucella antibodies were detected in pasteurized milk, this might be due to proper pasteurization, and is in accord with that reported by OIE (2005). Moreover this result is better result than that reported by Alves *et al.* (2001). The presence of positive antibodies for brucella in the raw milk samples might suggest infection and/or vaccination, as those herds followed regular vaccination programmes. Similarly negative results of antibiotic residues test were obtained, this may be due to proper follow up of antibiotic withdrawal periods which indicated the good quality of raw milk used. These results are in agreement with Van Schaik *et al.* (2002) and Yamaki *et al.* (2004). It is concluded that the values of chemical contents are within standards limits except for lactose, whose value was lower than the reported by the dairy plant. Low TBC for pasteurized milk was obtained, and the results of this study clearly illustrate that pasteurization plays an important role in the survival and destruction of different bacterial contaminants.

Hazard Analysis Critical Control Point—HACCP

Hazard Analysis Critical Control Point or HACCP is a systematic preventive approach to food safety and pharmaceutical safety that addresses physical, chemical, and biological hazards as a means of prevention rather than finished product inspection. HACCP is used in the food industry to identify potential food safety hazards, so that key actions can be taken to reduce or eliminate the risk of the hazards being realized.

The system is used at all stages of food production and preparation processes including packaging, distribution, etc. The Food and Drug Administration (FDA) and the United States Department of Agriculture (USDA) say that their mandatory HACCP programs for juice and meat are an effective approach to food safety and protecting public health. Meat HACCP systems are regulated by the USDA, while seafood and juice are regulated by the FDA. The use of HACCP is currently voluntary in other food industries.

A forerunner to HACCP was developed in the form of production process monitoring during World War II because traditional "end of the pipe" testing was not an efficient way to ferret out artillery shells that would not explode. HACCP itself was conceived in the 1960s when the US National Aeronautics and Space Administration (NASA) asked Pillsbury to design and manufacture the first foods for space flights.

Since then, HACCP has been recognized internationally as a logical tool for adapting traditional inspection methods to a modern, science-based, food safety system. Based on risk-assessment, HACCP plans allow both industry and government to allocate their resources efficiently in establishing and auditing safe food production practices. In 1994, the organization of *International HACCP Alliance* was established initially for the US meat and poultry industries to assist them with implementing HACCP and now its membership has been spread over other professional/industrial areas.

Hence, HACCP has been increasingly applied to industries other than food, such as cosmetics and pharmaceuticals. This method, which in effect seeks to plan out unsafe practices, differs from traditional "produce and test" quality control methods which are less successful and inappropriate for highly perishable foods. In the US, HACCP compliance is regulated by 21 CFR part 120 and 123. Similarly, FAO/WHO published a guideline for all governments to handle the issue in small and less developed food businesses.

History

On 4 October 1957, the Soviet Union launched Sputnik, the world's first satellite. American president Dwight D. Eisenhower responded by committing the United States to the space race. Eisenhower signed the National Aeronautics and Space Act on 29 July 1958 that created the National Aeronautics and Space Administration (NASA) to put an American satellite in orbit and to get a person in space.

Food played a critical part in the manned space program. The initial group involved in this were Herbert Hollander, Mary Klicka, and Hamed El-Bisi of the United States Army Laboratories in Natick, Massachusetts and Dr. Paul A. Lachance of the Manned Spaceflight Centre (Johnson Space Centre since February 1973) in Houston, Texas. Pillsbury joined the program as a contractor in 1959 with Howard E. Baumann representing the company as its lead scientist.

The main goal was to produce food that would not crumble under zero gravity, but also be safe to eat. Lachance imposed strict microbial requirements, including pathogen limits (including *E. coli*, *Salmonella*, and *Clostridium botulinum*) on all foods destined for space travel. All personnel involved realized that traditional quality control methods would be inadequate because there would be so much product testing involved for actual product to be used. NASA own requirements for Critical Control Points (CCP) in engineering management would be used as a guide for food safety.

CCP derived from Failure mode and effects analysis (FMEA) from NASA via the munitions industry to test weapon and engineering system reliability. Using that information, NASA and Pillsbury required contractors to identify "critical failure areas" and eliminate them from the system, a first in the food industry then. Baumann, a microbiologist by training, was so pleased with Pillsbury's experience in the space program that he advocated for his company to adopt what would become HACCP at Pillsbury.

Soon thereafter, Pillsbury was confronted with a food safety issue of its own when glass was found contaminated in farina, a cereal commonly used in infant food. Baumann's leadership promoted HACCP in Pillsbury for producing commercial foods, and applied to its own food production.

This led to a panel discussion at the 1971 National Conference on Food Protection that included examing CCPs and Good Manufacturing Practices in producing safe foods. Several botulism cases were attributed to under-processed low-acid canned foods in 1970-71. The United States Food and Drug Administration (FDA) asked Pillsbury to organize and conduct a training program on the inspection of canned foods for FDA inspectors. This 21 day program was first held in September 1972 with 11 days of classroom lecture and 10 days of canning plant evaluations. Canned food regulations (21 CFR 108, 21 CFR 110, 21 CFR 113, and 21 CFR 114) were first published in 1973. Pillsbury's training program to the FDA in 1972, titled "Food Safety through the Hazard Analysis and Critical Control Point System", was the first time that HACCP was used.

HACCP was initially set on three principles, now shown as principles one, two, and four in the section below. Pillsbury quickly adopted two more principles, numbers three and five, to its own company in 1975. It was further supported by the National Academy of Sciences (NAS) that governmental inspections by the FDA go from reviewing plant records to compliance with its HACCP system. A second proposal by the NAS led to the development of the National Advisory Committee on Microbiological Criteria for Foods (NACMCF) in 1987. NACMCF was initially responsible for defining HACCP's systems and guidelines for its application and were coordinated with the Codex Committee for Food Hygiene, that led to reports starting in 1992 and further harmonization in 1997. By 1997, the seven HACCP principles listed below became the standard. A year earlier, the American Society for Quality offered their first certifications for HACCP Auditors. (First known as Certified Quality Auditor-HACCP, they were changed to Certified HACCP Auditor (CHA) in 2004.

HACCP expanded in all realms of the food industry, going into meat, poultry, seafood, dairy, and has spread now from the farm to the fork.

The HACCP Seven Principles

Principle 1: Conduct a hazard analysis. - Plans determine the food safety hazards and identify the preventive measures the plan can apply to control these hazards. A food safety hazard is any biological, chemical, or physical property that may cause a food to be unsafe for human consumption.

Principle 2: Identify critical control points. - A Critical Control Point (CCP) is a point, step, or procedure in a food manufacturing process at which control can be applied and, as a result, a food safety hazard can be prevented, eliminated, or reduced to an acceptable level.

Principle 3: Establish critical limits for each critical control point. - A critical limit is the maximum or minimum value to which a physical, biological, or chemical hazard must be controlled at a critical control point to prevent, eliminate, or reduce to an acceptable level.

Principle 4: Establish critical control point monitoring requirements. - Monitoring activities are necessary to ensure that the process is under control at each critical control point. In the United States, the FSIS is requiring that each monitoring procedure and its frequency be listed in the HACCP plan.

Principle 5: Establish corrective actions. - These are actions to be taken when monitoring indicates a deviation from an established critical limit. The final rule requires a plant's HACCP plan to identify the corrective actions to be taken if a critical limit is not met. Corrective actions are intended to ensure that no product injurious to health or otherwise adulterated as a result of the deviation enters commerce.

Principle 6: Establish record keeping procedures. - The HACCP regulation requires that all plants maintain certain documents, including its hazard analysis and written HACCP plan, and records documenting the monitoring of critical control points, critical limits, verification activities, and the handling of processing deviations.

Principle 7: Establish procedures for ensuring the HACCP system is working as intended. - Validation ensures that the plants do what they were designed to do; that is, they are successful in ensuring the production of a safe product. Plants will be required to validate their own HACCP plans. FSIS will not approve HACCP plans in advance, but will review them for conformance with the final rule.

Verification ensures the HACCP plan is adequate, that is, working as intended. Verification procedures may include such activities as review of HACCP plans, CCP records, critical limits and microbial sampling and analysis. FSIS is requiring that the HACCP plan include verification tasks to be performed by plant personnel. Verification tasks would also be performed by FSIS inspectors. Both FSIS and industry will undertake microbial testing as one of several verification activities.

Verification also includes 'validation' - the process of finding evidence for the accuracy of the HACCP system (e.g. scientific evidence for critical limitations).

Standards

The seven HACCP principles are included in the international standard ISO 22000 FSMS 2005. This standard is a complete food safety and quality management system incorporating the elements of prerequisite programmes (GMP & SSOP), HACCP and the quality management system, which together form an organization's Total Quality Management system.

HACCP Training

HACCP management system trainings are only offered by several commercial enthusiasts. However, ASQ does provide Trained HACCP Auditor (CHA) exam to individuals seeking the professional training. In the UK the Chartered Institute of Environmental Health (CIEH) offer a HACCP for Food Manufacturing qualification accredited by the QCA (Qualifications and Curriculum Authority).

HACCP Application

Applied Range

It can apply to several food categories; sea food, bulk milk production line, Bulk Cream and Butter Production Line, animal meat industry, Organic Chemical Contaminants in Food, Corn Curl Manufacturing Plant and etc.

USA

- Fish and fishery products
- Fresh-cut produces
- Juice and nectary products
- Food outlets
- Meat and poultry products
- School food and services.

HACCP Implementation

It involves monitoring, verifying and validating of the daily work that is compliant with regulatory requirements in all stages all the time. The differences among those three types of work are given by Saskatchewan Agriculture and Food.

HACCP Versus ISO 22000

ISO 22000 is the new standard bound to replace HACCP on issues related to food safety. Although several companies, especially the big ones, have either implemented or are on the point of implementing ISO 22000, there are many others which are rather timid and/or reluctant to implement it. The main reason behind that is the lack of information and the fear that the new standard is too demanding in terms of bureaucratic work, from abstract of case study.

Emulsion

An emulsion is a mixture of two or more immiscible (unblendable) liquids. Emulsions are part of a more general class of two-phase systems of matter called colloids. Although the terms colloid and emulsion are sometimes used interchangeably, emulsion tends to imply that both the dispersed and the continuous phase are liquid. In an emulsion, one liquid (the dispersed phase) is dispersed in the other (the continuous phase).

Examples of emulsions include vinaigrettes, the photo-sensitive side of photographic film, milk and cutting fluid for metal working.

Structure and Properties of Emulsions

It is still common belief that emulsions basically do not display any structure, i.e., the droplets (or in case of dispersions, particles) dispersed in the liquid matrix (the "dispersion medium") are assumed to be statistically distributed. Therefore, for emulsions (like for dispersions) usually percolation theory is assumed to appropriately describe their properties.

However, percolation theory can only be applied if the system it should describe is in or close to thermodynamic equilibrium. There are very few studies about the structure of emulsions (dispersions), although they are plentiful in type and in use all over the world in innumerable applications.

In the following, only such emulsions will be discussed with a dispersed phase diameter of less than 1 μm. To understand the formation and properties of such emulsions (including dispersions), it must be considered, that the dispersed phase exhibits a "surface," which is covered ("wet") by a different "surface" which hence are forming an interface (chemistry). Both surfaces have to be created (which requires a huge amount of energy), and the interfacial tension (difference of surface tension) is not compensating the energy input, if at all.

A review article in introduces into various attempts to describe dispersions / emulsions. Dispersion is a process by which (in the case of solids becoming dispersed in a liquid) agglomerated particles are separated from each other and a new interface, between an inner surface of the liquid dispersion medium and the surface of the particles to be dispersed, is generated. Dispersion is a much more complicated (and less well understood) process than most people believe.

The above cited review article also displays experimental evidence for that dispersions have a structure very much different from any kind of statistical distribution (which would be characteristic for a system in thermodynamic equilibrium, but in contrast very much showing structures similar to self-organisation which can be described by non-equilibrium thermodynamics. This is the reason why some liquid dispersions turn to become gels or even solid at a concentration of a dispersed phase above a certain critical concentration (which is dependant on particle size and interfacial tension). Also the sudden appearance of conductivity in a system of a dispersed conductive phase in an insulating matrix has been explained. The above cited review article also introduces into some first complete non-equilibrium thermodynamics theory of dispersions.

Appearance and Properties

Emulsions are made up of a dispersed and a continuous phase;

the boundary between these phases is called the interface. Emulsions tend to have a cloudy appearance, because the many phase interfaces scatter light that passes through the emulsion. Emulsions are unstable and thus do not form spontaneously. The basic colour of emulsions is white. If the emulsion is dilute, the Tyndall effect will scatter the light and distort the colour to blue; if it is concentrated, the colour will be distorted towards yellow. This phenomenon is easily observable on comparing skimmed milk (with no or little fat) to cream (high concentration of milk fat). Microemulsions and nanoemulsions tend to appear clear due to the small size of the disperse phase.

Energy input through shaking, stirring, homogenizing, or spray processes are needed to initially form an emulsion. Over time, emulsions tend to revert to the stable state of the phases comprising the emulsion; an example of this is seen in the separation of the oil and vinegar components of Vinaigrette, an unstable emulsion that will quickly separate unless shaken continuously.

Whether an emulsion turns into a water-in-oil emulsion or an oil-in-water emulsion depends on the volume fraction of both phases and on the type of emulsifier. Generally, the Bancroft rule applies: emulsifiers and emulsifying particles tend to promote dispersion of the phase in which they do not dissolve very well; for example, proteins dissolve better in water than in oil and so tend to form oil-in-water emulsions (that is they promote the dispersion of oil droplets throughout a continuous phase of water).

Instability

There are three types of instability: flocculation, creaming, and coalescence. Flocculation describes the process by which the dispersed phase comes out of suspension in flakes. Coalescence is another form of instability, which describes when small droplets combine to form progressively larger ones. Emulsions can also undergo creaming, the migration of one of the substances to the top (or the bottom, depending on the relative densities of the two phases) of the emulsion under the influence of buoyancy or centripetal force when a centrifuge is used.

Surface active substances (surfactants) can increase the kinetic stability of emulsions greatly so that, once formed, the emulsion does not change significantly over years of storage. A Non-Ionic surfactant solution can become self-contained under the force of its own surface tension, remaining in the shape of its previous container for some time after the container is removed. Superfluids flow with zero friction and can escape their containers; an ionic solution tends to retain its current shape. "Emulsion stability refers to the ability of an emulsion to resist change in its properties over time." D.J. McClements.

Technique Monitoring Physical Stability

Multiple light scattering coupled with vertical scanning is the most widely used technique to monitor the dispersion state of a product, hence identifying and quantifying destabilisation phenomena. It works on concentrated emulsions without dilution. When light is sent through the sample, it is backscattered by the droplets. The backscattering intensity is directly proportional to the size and volume fraction of the dispersed phase. Therefore, local changes in concentration (Creaming) and global changes in size (flocculation, coalescence) are detected and monitored.

Accelerating Methods for Shelf Life Prediction

The kinetic process of destabilisation can be rather long (up to several months or even years for some products) and it is often required for the formulator to use further accelerating methods in order to reach reasonable development time for new product design. Thermal methods are the most commonly used and consists in increasing temperature to accelerate destabilisation (below critical temperatures of phase inversion or chemical degradation). Temperature affects not only the viscosity, but also interfacial tension in the case of non-ionic surfactants or more generally interactions forces inside the system. Storing a dispersion at high temperatures enables to simulate real life conditions for a product (e.g. tube of sunscreen cream in a car in the summer), but also to accelerate destabilisation processes up to 200 times. Mechanical acceleration including vibration, centrifugation and agitation are sometimes used. They subject the product to different forces that pushes the droplets against one another, hence helping in the film drainage. However, some emulsions would never coalesce in normal gravity, while they do under artificial gravity. Moreover segregation of different populations of particles have been highlighted when using centrifugation and vibration.

Emulsifier

An emulsifier (also known as an emulgent) is a substance which stabilizes an emulsion by increasing its kinetic stability. One class of emulsifiers is known as surface active substances, or surfactants. Examples of food emulsifiers are egg yolk (where the main emulsifying agent is lecithin), honey, and mustard, where a variety of chemicals in the mucilage surrounding the seed hull act as emulsifiers; proteins and low-molecular weight emulsifiers are common as well. Soy lecithin is another emulsifier and thickener. In some cases, particles can stabilize emulsions as well through a mechanism called Pickering stabilization. Both mayonnaise and Hollandaise sauce are oil-in-water emulsions that are stabilized with egg yolk lecithin or other types of food additives such as Sodium stearoyl lactylate.

Detergents are another class of surfactant, and will physically interact with both oil and water, thus stabilizing the interface between oil or water droplets in suspension. This principle is exploited in soap to remove grease for the purpose of cleaning. A wide variety of emulsifiers are used in pharmacy to prepare emulsions such as creams and lotions. Common examples include emulsifying wax, cetearyl alcohol, polysorbate 20, and ceteareth 20. Sometimes the inner phase itself can act as an emulsifier, and the result is nanoemulsion - the inner state disperses into nano-size droplets within the outer phase. A well-known example of this phenomenon, the ouzo effect, happens when water is poured in a strong alcoholic anise-based beverage, such as ouzo, pastis, arak or raki. The anisolic compounds, which are soluble in ethanol, now form nano-sized droplets and emulgate within the water. The colour of such diluted drink is opaque and milky.

In Food

Oil-in-water emulsions are common in food. Notable examples include:

- Crema in espresso – coffee oil in water (brewed coffee), unstable
- Hollandaise sauce – similar to mayonnaise
- Mayonnaise – vegetable oil in lemon juice or vinegar, with egg yolk lecithin as emulsifier
- Vinaigrette – vegetable oil in vinegar; if prepared with only oil and vinegar (without an emulsifier), yields an unstable emulsion.

In Medicine

In pharmaceutics, hairstyling, personal hygiene and cosmetics, emulsions are frequently used. These are usually oil and water emulsions, but which is dispersed and which is continuous depends on the pharmaceutical formulation. These emulsions may be called creams, ointments, liniments (balms), pastes, films or liquids, depending mostly on their oil and water proportions and their route of administration.

The first 5 are topical dosage forms, and may be used on the surface of the skin, transdermally, ophthalmically, rectally or vaginally. A very liquidy emulsion may also be used orally, or it may be injected using various routes (typically intravenously or intramuscularly). Popular medicated emulsions include calamine lotion, cod liver oil, Polysporin, cortisol cream, Canesten and Fleet.

Microemulsions are used to deliver vaccines and kill microbes. Typically, the emulsions used in these techniques are nanoemulsions of soybean oil, with particles that are 400-600 nm in diameter. The process is not chemical, as with other types of antimicrobial treatments, but mechanical. The smaller the droplet, the greater the surface tension and thus the greater the force to merge with other lipids.

The oil is emulsified using a high shear mixer with detergents to stabilize the emulsion, so when they encounter the lipids in the membrane or envelope of bacteria or viruses, they force the lipids to merge with themselves. On a mass scale, this effectively disintegrates the membrane and kills the pathogen. This soybean oil emulsion does not harm normal human cells nor the cells of most other higher organisms. The exceptions are sperm cells and blood cells, which are vulnerable to nanoemulsions due to their membrane structures. For this reason, these nanoemulsions are not currently used intravenously. The most effective application of this type of nanoemulsion is for the disinfection of surfaces. Some types of nanoemulsions have been shown to effectively destroy HIV-1 and various tuberculosis pathogens, for example, on non-porous surfaces.

In Fire Fighting

Emulsifying agents are effective at extinguishing fires on small thin layer spills of flammable liquids (Class B fires). Extinguishment is achieved by encapsulating the fuel in a fuel-water emulsion thereby trapping the flammable vapors in the water phase. This emulsion is achieved by applying an aqueous surfactant solution to the fuel through a high pressure nozzle. Emulsifiers are not effective at extinguishing large Class B fuel in depth fires. This is because the amount of agent needed for extinguishment is a function of the volume of the fuel whereas agents such as aqueous film forming foam (AFFF) need only cover the surface of the fuel to achieve vapor mitigation.

Uses

Emulsions are mainly used in many major chemical industries. In the pharmaceutical industry they are used to make medicines with a more appealing flavour and to improve value by controlling the amount of active ingredients. The most widely-used emulsions are non-ionic because they have low toxicity, but cationic emulsions are also used in some products because of their antimicrobial properties. Emulsions are also used in making many hair and skin products, such as various types of oils and waxes.

Shelf-life Predicting Methods for Milk

New analysis allows prediction of shelf life for pasteurized bottled milk. At I&A Lab we think that the most important part of a plant is production, and the laboratory should provide help to production to solve and prevent the development of quality problems. Production faces different types of problems like:

1- Quality of the raw milk received

2- Equipment

3- CIP

4- Packing material

5- Personnel.

All of these can affect the quality of the final product and as a consequence the shelf life.

Our analysis is designed for use with pasteurized milk. To establish the shelf life of bottled milk it is necessary to be able to predict how the product is going to behave in the stores, at the correct temperature, and to know what is going to happen if the product is abused. The ideal is to have this prediction in a reasonable amount of time, the sooner after the milk is bottled being better.

There are different groups of microorganisms capable of growing and spoiling the milk, for our study we consider that the mesophilic (growing between 68° and 113°F), and psychrophilic (growing between 20° and 68°F) are the most important groups, so we focus our effort on the mesophilic's which can spoil the milk when it is abused, and the psychrophilic's which can spoil the milk even at temperatures below 45°F. Most psychrophilic microorganisms are a result of post - pasteurization contamination, due to the fact that they usually die with pasteurization. It is known that milk is an excellent media for the growth of many microorganisms, it is also known that if the milk is kept under 45°F many microorganisms stop or decrease their growth.

The legal analysis for pasteurized bottled milk uses one milliliter of milk in a solid media for total aerobic counts and coliform counts.

It is not a requirement to analyze psychrophilics, but the most widely used analysis takes 7 to 10 days to complete. The solid media employed is very different from liquid milk, so much so that it takes 48 hrs to see colonies growing in the solid media plates, incubated at 113°F, while it takes only a few hours for the milk to spoil. This means that for most of the microorganisms it is easier to grow in a liquid than in a solid.

The reason for the use of solid media is to be able to count the amount of colonies per milliliter of milk. We have developed a new method using liquid milk in a volume 10 times higher than normal in special equipment using new software. This allows us to detect microorganisms in very low concentrations. We also employ two different medias for psychrophilic bacteria, along with media for pseudomonas, and mesophilic bacteria. This allows us to have a wide spectrum of detectable milk inhabiting microorganisms.

We created a database with our results and compared them against the results of the plated fresh samples, and the same samples preincubated at 68°F for 16 hrs. Finally, the milk was flavoured over a period of time until it was determined to be no longer acceptable for consumption. After more than 8000 samples being analyzed we are able to detect in 20 hrs samples which will have 10 days or less of shelf life. If the milk is contaminated for any reason we will be able to give the production plant this information. We are able to report in 24 hrs if the milk has any problem which may compromise its shelf life.

We will issue a second report in 48 hrs stating the amount of psychrophilic, pseudomonas, coliform, and mesophilic bacteria in the milk sample, with a prediction, in number of days, with which the milk will be in good condition if stored at 45°F. Also, the number of days of shelf life if the milk is kept between 50° and 55°F will be supplied. During our study we found a type of bacteria capable of growing at 68°F. We identified this as Bacillus megaterium (mesophilic). As its presence occurred frequently we studied it and determined that even though it is capable of growing at room temperatures it is not capable of growing below 45°F, however it grows rapidly if the milk is abused. As the milk is a biological product, and there are millions of different microorganisms capable of growing and making changes in the milk, we will continue our study. In the future we will be able to provide more information to help production managers make better decisions regarding the shelf life of their products.

Determination of the End of Shelf Life for Milk

Using Weibull Hazard Method

Undesirable changes in dairy products may be instigated by microbial growth and metabolism or by chemical reactions. The determinants of shelf life of fresh dairy products are usually the spoilage bacteria that have the ability to grow at refrigerated temperatures. This microbial growth induces changes in the taste and odour of milk such as sour, putrid, bitter, malty, fruity, rancid and unclean. In addition, psychrotrophs which are common contaminants in milk, synthesize enzymes, many of which survive the pasteurization heat treatment and during storage biochemically alter the milk and eventually cause spoilage.

Growth of psychrotrophic bacteria is predominantly responsible for influencing the keeping quality of milk and dairy products held below 7°C. Raw and pasteurized milk usually spoils when held at refrigeration temperatures because of the effects of psychrotrophic contaminates.

The populations of microorganisms needed to cause detectable changes in milk varies among genera and species within a genus, but levels at which flavour changes occur are similar at 6 and 20°C. Milk spoilage by psychrotrophs was reported in the range of populations of 1 x 102 to 1 x 109 per ml.

It is therefore unclear whether pschrotrophs counts can be used as an index in the determination of milk quality or shelf life from a sensory standpoint. As noted earlier, microbial spoilage leads to sensory deterioration of the milk. It may therefore be suggested that the microbial quality of the milk should correlate well to its sensory end of shelf life. The end of shelf life can be determined from sensory data by various graphical methods. The use of hazard rate for shelf life testing of food was introduced by Gacula (1975). Using this method, one can determine the end of shelf life according to the percent of customers a company is prepared to displease.

The maximum likelihood graphical procedure, or Weibull Hazard method has been used for shelf life of luncheon meats, oat bran cereal, ice cream, cottage cheese, Bockwurst sausages and butter, and other food products. The objective of this research was to determine whether or not a consumer determined end of sensory shelf life can be described by some microbial index regardless of the temperature conditions that the milk is stored at.

Materials and Methods

Milk

The milk used in this study was TLC⁼ fat free milk with added Calcium. This milk is also fortified with nonfat milk proteins. The raw milk was held for no longer than 48 hours at 2°C before processing and then pasteurized (20 s, 80°C). Half gallon, paper board cartons of milk were picked up within 2 h after bottling, taken off the production line consecutively by a plant supervisor, in order to minimize variability between cartons. The cartons of milk were transported to the University of Minnesota on ice in a cooler. Immediately after arrival, the milk was sampled for microbial quality and tasted by three expert dairy panelists to ensure that the milk was of good quality.

Microbial Counts

Total aerobic bacteria as well as psychrotrophic bacteria were enumerated using 3M Petrifilm (3M Co., St. Paul, MN). Samples were diluted in 0.1% peptone, and 1 ml of sample was transfered onto the film in duplicate. The Petrifilm contained standard method nutrients and a cold water soluble gelling agent (8, 20). The bottom film is coated with nutrients and gelling agent, while the top film is coated with the gelling agent and 2,3, 5-triphenylterazoluim chloride (TTC). Colonies appear red and were counted following incubation at 37°C for 48 h for total aerobic or 10 d at 7°C for pyschrotrophs.

Microbial Growth

The growth of total aerobic bacteria and psychrotrophic bacteria in the milk was measured at five constant temperatures: 2, 5, 7, 12, and 15°C (± 1°C). A TempTale (Sensitech, Beverly, MA) temperature recorder, placed in the coolers along with the milk verified the temperature history.

Samples were drawn at predetermined intervals according to the storage temperature conditions. The lag times were determined graphically, and the growth rate constant was calculated through linear regression of the exponential phase of the growth curves.

Sensory Testing

The sensory testing was carried out following the Weibull Hazard method , where the initial number of panelists was $n0 = 3$ and the constant with which the number of panelists was increased for each subsequent test was $nc = 1$. The interval between sensory testing was predetermined for each of the five different storage conditions.

Because the spoilage of milk at 14°C occurred at an accelerated rate, sensory samples were held overnight so that the sensory test could be carried out at a convenient time for the panelists. The panelists were prescreened and were required to meet the criterion that they consume at least one 8-ounce glass of milk a day.

A pool of 33 panelists who met this requirement, 16 male and 17 female, ranging in age form 18 to 45 were available for sensory testing. The panelists were financially compensated according to the number of samples that they tested.

For each sensory test a sample of milk was taken from the milk cartons, poured into a glass flask and the flask was immediately placed into an ice bath to slow down any microbial growth that might have caused any further deterioration of the sensory quality of the milk. Approximately 10 ml of milk was poured into cups that were labelled with random three digit numbers for identification purposes.

A tray of milk samples for each of the panelists was prepared approximately half an hour to an hour before sensory testing took place.

To ensure that the samples were all the same temperature when the panelists received their trays the trays were stored in conventional home refrigerators held at 4°C. The trays, consisted of samples of milk from the different storage temperatures.

The trays were presented to the panelists in sensory booths where the sensory testing was held. Panelists were asked to taste the first sample and determine whether the milk was acceptable or unacceptable, where a response of acceptable implied that the panelist would be willing to drink an entire glass of the sample. Panelists were asked to wait two minutes between samples and to rinse their mouths with water in between.

The end of sensory shelf life was determined at 69.3 % cumulative hazard or a critical probability of 50%. The data was regressed using the least squares method up to 100% cumulative hazard.

Results and Discussion

Microbial Growth

The growth of the total aerobic bacteria and the psychrotrophic bacteria were obtained at 2, 5, 7, 12, and 14°C (± 1°C).

The total aerobic microbial population exhibited typical growth curves at all temperatures, however, at 2 and 5°C the milk reached the end of sensory shelf life before the end of the lag phase. The psychrotrophic bacteria, exhibit distinct lag and log growth phases at 5, 7, 12, and 14°C. At 2°C however, a rapid growth of psychrotrophs occurred immediately following the opening of the milk carton for sampling, thus, obtaining a growth curve for psychrotrophs at 2°C was not practical within this experimental setup.

It is likely that the rapid growth following opening of the milk carton was due to the change in available oxygen.

Once the carton is opened the amount of available oxygen for the microorganisms in the milk increases and facilitates their growth. Sinclair and Stokes (1963) support this explanation with the discovery that in general, due to an increase in the availability of oxygen higher counts are observed.

The growth of both aerobic and pschrotrophs populations at 5°C did not exhibit a distinct logarithmic phase. The absence of a logarithmic growth phase at 5°C can be attributed to the fact that the samples were taken from more than one carton throughout the experiment.

Thus the variability in the population of microorganisms from carton to carton may result in difficulties in detecting a distinct lag and exponential phase. The cartons were taken from the production line in consecutive order so that the cartons could be considered to be identical and so the sampling from the cartons during the growth curve study could be made randomly between the opened cartons.

However, Maxcy and Wallen (1983) found that heterogeneity between cartons was apparent even when samples were taken sequentially from a single production line.

Growth Parameters

It was not possible to determine the duration of the lag phase for psychrotrophic bacteria of the milk stored at 2°C because the psychrotrophic counts showed no distinctive pattern.

Since the milk stored at 5° and 2°C reached the sensory end point during the lag phase of both the total aerobic bacteria and psychrotrophic bacteria, the exponential growth rates for the bacteria at these two storage temperatures, were not calculated. Data presented by Fu (1989) showed that the temperature dependence of the growth rate constants and the lag time for microbial growth in a model milk system fits the Arrhenius model.

Indigenous (Indian) Dairy Products

A variety of dairy projects are indigenous to India and an important part of Indian cuisine. The majority of these products can be broadly classified into curdled products, like chhena, or non-curdled products, like khoa.

Curdled Dairy Products

- Paneer is an unaged, acid-set, non-melting farmer cheese made by curdling heated milk with lemon juice or other non-rennet food acid, and then removing the whey and pressing the result into a dry unit.

- Chhena is like paneer, except some whey is left and the mixture is beaten thoroughly until it becomes soft, of smooth consistency, and malleable but firm.

- Sandesh is a confection made from chhena mixed with sugar then grilled lightly to caramelize, but removed from heat and molded into a ball or some shape.

- Rasgulla is a confection made from mixture of chhena and semolina rolled into a ball and boiled in syrup.

Non-curdled Dairy Products

- Khoa or Mawa is made by reducing milk in an open pan over heat.

- Peda is a confection made by mixing sugar with khoa and adding flavoring, such as cardamom.

- Barfi is a confection made by reducing milk and sugar until it solidifies and adding flavoring, such as pistachio.

- Gulab jamun is a confection made by mixing khoa and sugar, caramelizing it by frying, and soaking it in syrup containing rosewater.

- Kulfi is made from slowly freezing sweetened condensed milk. In comparison to ice cream, kulfi is not whipped or otherwise aerated.

- Ghee is type of clarified butter that is cooked long enough to caramelize the milk sugar and sterilize the liquid.

Fermented Dairy Products

- Mishti doi is *dahi* (Indian yogurt) mixed with sugar
- Shrikhand is strained yoghurt mixed with sugar, and often flavourings such as cardamom, saffron, or fruit.
- Wheyvit is an alcoholic beverage prepared by fermenting whey with yeast.

Other Dairy Products

- Kheer is made by boiling rice or broken wheat with milk and sugar, and sometimes flavoured with cardamom, raisins, saffron, pistachios, or almonds.
- Chhena Murki is made by frying cubes of chhena to burn the outside, then soaking them in syrup flavoured with cardamom.
- Pantooa is like gulab jamun, except with some chhena mixed with the usual ingredients.

Chapter 3

Milk Chemistry—An Introduction

Physical Status of Milk

About 87% of milk is water, in which the other constituents are distributed in various forms. We distinguish among several kinds of distribution according to the type and size of particle present in the liquid.

Kind of solution	Particle diameter (nm)
Ionic solution	0.01–1
Molecular solution	0.1–1
Colloid (fine dispersion)	1–100
Coarse dispersion (suspension or emulsion)	50–100

In milk we find examples of emulsions, colloids, molecular and ionic solutions.

Ionic Solutions

An ionic solution is obtained when the forces that hold the ions together in a solid salt are overcome. The dissolved salt breaks up into ions which float freely in the solvent. Thus when common salt—sodium chloride—is dissolved in water it becomes an ionic solution of free sodium and chloride ions. Ionic solutions are largely of inorganic compounds.

Molecular Solutions

In a molecular solution the molecules are only partly, if at all, dissociated into ions.

The degree of dissociation represents an equilibrium which is influenced by other substances in the solution and by the pH (or hydrogen ion concentration) of the solution. Molecular solutions are usually of organic compounds.

Colloids

In a colloid, one substance is dispersed in another in a finer state than an emulsion but the particle size is larger than that in a true solution. Colloidal systems are classified according to the physical state of the two phases. In a colloid, solid particles consisting of groups of molecules float freely. The particles in a colloid are much smaller than those in a suspension and a colloid is much more stable.

Emulsions

An emulsion consists of one immiscible liquid dispersed in another in the form of droplets—the disperse phase. The other phase is referred to as the continuous phase. The systems have minimal stability and require the presence of a surface-active or emulsifying agent for stability. In foods, emulsions usually contain oil and water. If water is the continuous phase and oil the disperse phase, it is an oil-in-water (o/w) emulsion, e.g. milk or cream. In the reverse case the emulsion is a water-in-oil (w/o) type, e.g. butter. In summary, an emulsion consists of three elements, the continuous phase, the disperse phase and the emulsifying agent.

Dispersions

A dispersion is obtained when particles of a substance are dispersed in a liquid. A suspension consists of solid particles dispersed in a liquid, and the force of gravity can cause them to sink to the bottom or float to the top. For example, fine sand, dispersed in water, soon settles out.

PH and Acidity

An acid is a substance which dissociates to produce hydrogen ions in solution. A base (alkaline) is a substance which produces hydroxyl ions in solution. It can equally be stated that an acid is a substance which donates a proton and a base is a substance which accepts a proton. The symbol pH is used to denote acidity; it is inversely related to hydrogen ion concentration.

Neutrality is pH 7

Acidity is less than pH 7

Alkalinity is more than pH 7

Fresh milk has a pH of 6.7 and is therefore slightly acidic.

When an acid is mixed with a base, neutralisation takes place; similarly a base will be neutralised by an acid.

Buffer Solutions

Buffers are defined as materials that resist a change in pH on addition of acid or alkali. Characteristically they consist of a weak acid or a weak base and its salt. Milk contains a large number of these substances and consequently behaves as a buffer solution. Fresh cows milk has a pH of between 6.7 and 6.5. Values higher than 6.7 denote mastitic milk and values below pH 6.5 denote the presence of colostrum or bacterial deterioration. Because milk is a buffer solution, considerable acid development may occur before the pH changes. A pH lower than 6.5 therefore indicates that considerable acid development has taken place. This is normally due to bacterial activity.

Litmus test papers, which indicate pH, are used to test milk activity; pH measurements are often used as acceptance tests for milk.

Measuring milk acidity is an important test used to determine milk quality. Acidity measurements are also used to monitor processes such as cheese-making and yoghurt-making. The titratable acidity of fresh milk is expressed in terms of percentage lactic acid, because lactic acid is the principal acid produced by fermentation after milk is drawn from the udder and fresh milk contains only traces of lactic acid. However, due to the buffering capacity of the proteins and milk salts, fresh milk normally exhibits an initial acidity of 0.14 to 0.16% when titrated using sodium hydroxide to a phenolphthalein end-point.

Milk Constituents

The quantities of the main milk constituents can vary considerably depending on the individual animal, its breed, stage of lactation, age and health status. Herd management practices and environmental conditions also influence milk composition. The average composition of cows milk is shown in Table 1.

Table 1: Composition of cows milk

Main constituent	Range(%)	Mean(%)
Water	85.5 – 89.5	87.0
Total solids	10.5 – 14.5	13.0
Fat	2.5 – 6.0	4.0
Proteins	2.9 – 5.0	3.4
Lactose	3.6 – 5.5	4.8
Minerals	0.6 – 0.9	0.8

Water is the main constituent of milk and much milk processing is designed to remove water from milk or reduce the moisture content of the product.

Milk Fat

If milk is left to stand, a layer of cream forms on the surface. The cream differs considerably in appearance from the lower layer of skim milk.

Under the microscope cream can be seen to consist of a large number of spheres of varying sizes floating in the milk. Each sphere is surrounded by a thin skin—the fat globule membrane—which acts as the emulsifying agent for the fat suspended in milk. The membrane protects the fat from enzymes and prevents the globules coalescing into butter grains. The fat is present as an oil-in-water emulsion: this emulsion can be broken by mechanical action such as shaking.

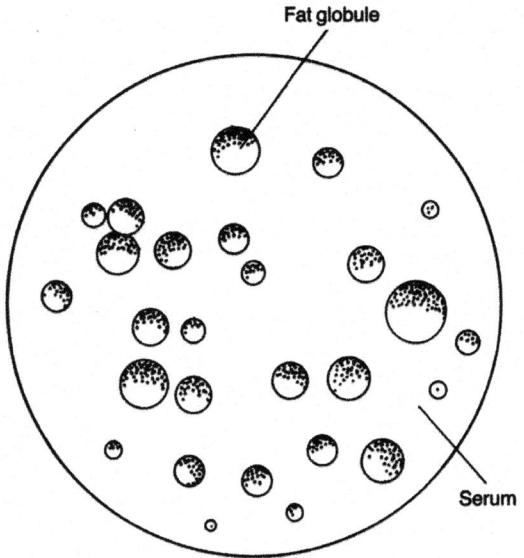

Figure 1. Fat globules in milk.

Fats are partly solid at room temperature. The term oil is reserved for fats that are completely liquid at room temperature. Fats and oils are soluble in non-polar solvents, e.g. ether.

About 98% of milk fat is a mixture of triacyl glycerides. There are also neutral lipids, fat-soluble vitamins and pigments (e.g. carotene, which gives butter its yellow colour), sterols and waxes. Fats supply the body with a concentrated source of energy: oxidation of fat in the body yields 9 calories/g. Milk fat acts as a solvent for the fat-soluble vitamins A, D, E and K and also supplies essential fatty acids (linoleic, linolenic and arachidonic).

A fatty-acid molecule comprises a hydrocarbon chain and a carboxyl group (-COOH). In saturated fatty acids the carbon atoms are linked in a chain by single bonds. In unsaturated fatty acids there is one double bond and in poly-unsaturated fatty acids there is more than one double bond. Examples of each type of fatty acid are shown in Figure.

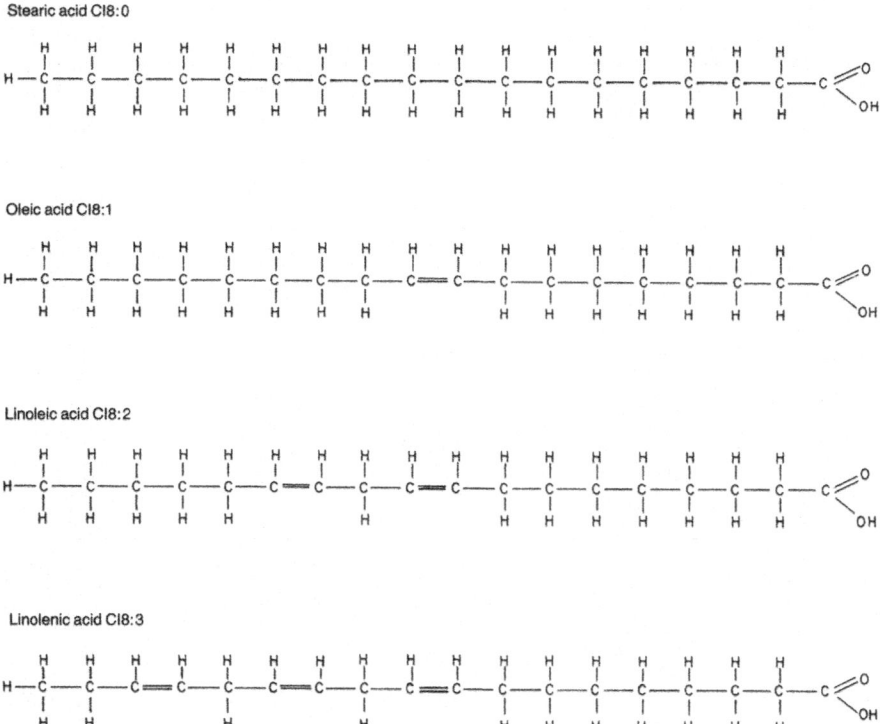

Figure 2. Structural formulae of four 18-carbon fatty acids varying in degree of saturation.

Fatty acids vary in chain length from 4 carbon atoms, as in butyric acid (found only in butterfat), to 20 carbon atoms, as in arachidonic acid. Nearly all the fatty acids in milk contain an even number of carbon atoms.

Fatty acids can also vary in degree of unsaturation, e.g. C18:0 stearic (saturated), C18:1 oleic (one double bond), C18:2 linoleic (two double bonds), C18:3 linolenic (three double bonds).

The most important fatty acids found in milk triglycerides are shown in Table 2. Fatty acids are esterified with glycerol as follows:

$$
\begin{array}{ccc}
H_2-C-OH & HOOC-R_1 & H_2-C-OOCR_1 \\
| & & | \\
H-C-OH & + \quad HOOC-R_2 \rightarrow & H-C-OOCR_2 + 3H \\
| & & | \\
H_2-C-OH & HOOC-R_3 & H_2-C-OOCR_3
\end{array}
$$

Glycerol + fatty acids '! triglyceride (fat) + water

Table 2. Principal fatty acids found in milk triglycerides.

	Molecular formula	*Chain length*	*Melting point*
Butyric	$CH_3(CH_2)_{2COOH}$	C_4	$-8°C$
Caproic	$CH_3(CH_2)_{4COOH}$	C_6	$-2°C$
Caprylic	$CH_2(CH_2)_{6COOH}$	C_8	$16°C$
Capric	$CH_3(CH_2)_{8COOH}$	C_{10}	$31.5°C$
Lauric	$CH_3(CH_2)_{10COOH}$	C_{12}	$44°C$
Myristic	$CH_3(CH_2)_{12COOH}$	C_{14}	$58°C$
Palmitic	$CH_3(CH_2)_{14COOH}$	C_{16}	$64°C$
Stearic	$CH_3(CH_2)_{16COOH}$	C_{18}	$70°C$
Arichidonic	$CH_3(CH_2)_{18COOH}$	C_{20}	
Oleic	$CH_3(CH_2)_{7CH}=CH(CH_2)_{7COOH}$	$C_{18:1}$	$13°C$
Linoleic	$CH_3(CH_2)_4(CH=CH.CH_2)_2(CH_2)_{6COOH}$	$C_{18:2}$	$-5°C$
Linolenic	$CH_3.CH_2(CH=CH.CH_2)_3(CH_2)_{6COOH}$	$C_{18:3}$	

The melting point and hardness of the fatty acid is affected by:

· the length of the carbon chain, and

· the degree of unsaturation.

As chain length increases, melting point increases. As the degree of unsaturation increases, the melting point decreases.

Fats composed of short-chain, unsaturated fatty acids have low melting points and are liquid at room temperature, i.e. oils. Fats high in long-chain saturated fatty acids have high melting points and are solid at room temperature. Butterfat is a mixture of fatty acids with different melting points, and therefore does not have a distinct melting point. Since butterfat melts gradually over the temperature range of 0–40°C, some of the fat is liquid and some solid at temperatures between 16 and 25°C. The ratio of solid to liquid fat at the time of churning influences the rate of churning and the yield and quality of butter. Fats readily absorb flavours. For example, butter made in a smoked gourd has a smokey flavour. Fats in foods are subject to two types of deterioration that affect the flavour of food products.

1. *Hydrolytic rancidity:* In hydrolytic rancidity, fatty acids are broken off from the glycerol molecule by lipase enzymes produced by milk bacteria. The resulting free fatty acids are volatile and contribute significantly to the flavour of the product.

2. *Oxidative rancidity:* Oxidative rancidity occurs when fatty acids are oxidised. In milk products it causes tallowy flavours. Oxidative rancidity of dry butterfat causes off-flavours in recombined milk.

Milk Proteins

Proteins are an extremely important class of naturally occurring compounds that are essential to all life processes. They perform a variety of functions in living organisms ranging from providing structure to reproduction. Milk proteins represent one of the greatest contributions of milk to human nutrition. Proteins are polymers of amino acids. Only 20 different amino acids occur, regularly in proteins. They have the general structure:

$$
\begin{array}{c}
NH \\
| \\
R - CH - COOH \\
| \\
H
\end{array}
$$

R represents the organic radical. Each amino acid has a different radical and this affects the properties of the acid. The content and sequence of amino acids in a protein therefore affect its properties. Some proteins contain substances other than amino acids, e.g. lipoproteins contain fat and protein. Such proteins are called conjugated proteins:

Phosphoproteins: Phosphate is linked chemically to these proteins—examples include casein in milk and phosphoproteins in egg yolk.

Lipoproteins: These combinations of lipid and protein are excellent emulsifying agents. Lipoproteins are found in milk and egg yolk.

Chromoproteins: These are proteins with a coloured prosthetic group and include haemoglobin and myoglobin.

Casein

Casein was first separated from milk in 1830, by adding acid to milk, thus establishing its existence as a distinct protein. In 1895 the whey proteins were separated into globulin and albumin fractions.

It was subsequently shown that casein is made up of a number of fractions and is therefore heterogeneous. The whey proteins are also made up of a number of distinct proteins as shown in the scheme in Figure.

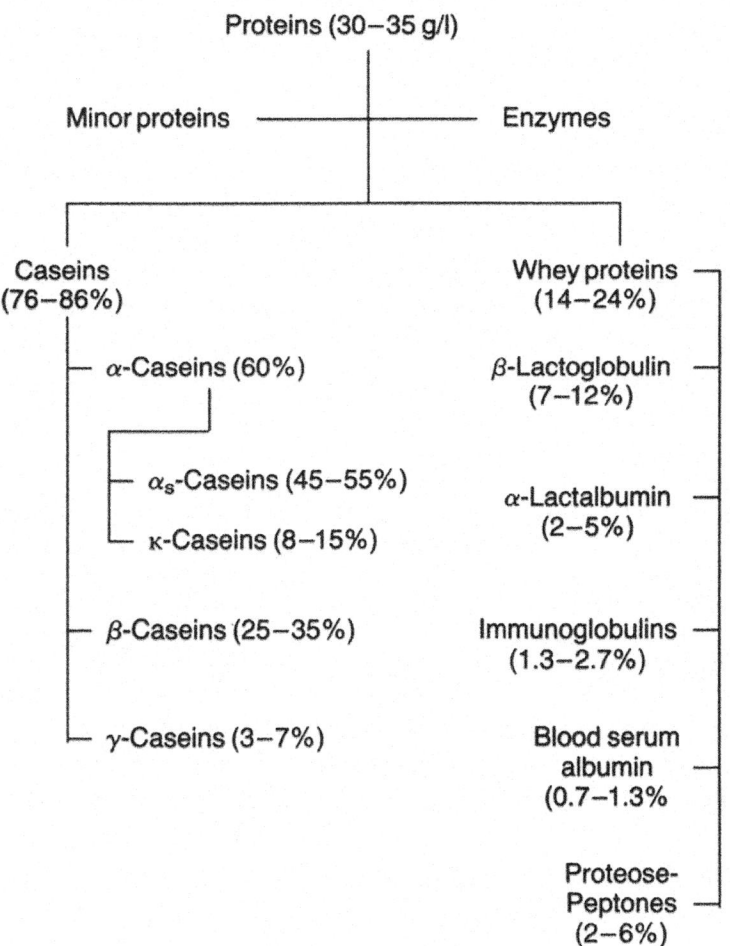

Figure 3. Milk protein fractions.

Casein is easily separated from milk, either by acid precipitation or by adding rennin. In cheese-making most of the casein is recovered with the milk fat. Casein can also be recovered from skim milk as a separate product. Casein is dispersed in milk in the form of micelles. The micelles are stabilised by the K-casein. Caseins are hydrophobic but K-casein contains a hydrophilic portion known as the glycomacropeptide and it is this that stabilises the micelles. The structure of the micelles is not fully understood. When the pH of milk is changed, the acidic or basic groups of the proteins will be neutralised. At the pH at which the positive charge on a protein equals exactly the negative charge, the net total charge of the protein is zero. This pH is called the isoelectric point of the protein (pH 4.6 for casein). If an acid is added to milk, or if acid-producing bacteria are allowed to grow in milk, the pH falls. As the pH falls the charge on casein falls and it precipitates. Hence milk curdles as it sours, or the casein precipitates more completely at low pH.

Whey Proteins

After the fat and casein have been removed from milk, one is left with whey, which contains the soluble milk salts, milk sugar and the remainder of the milk proteins. Like the proteins in eggs, whey proteins can be coagulated by heat. When coagulated, they can be recovered with caseins in the manufacture of acid-type cheeses. The whey proteins are made up of a number of distinct proteins, the most important of which are β-lactoglobulin and lactoglobulin. β-lactoglobulin accounts for about 50% of the whey proteins, and has a high content of essential amino acids. It forms a complex with K-casein when milk is heated to more than 75°C, and this complex affects the functional properties of milk. Denaturation of β-lactoglobulin causes the cooked flavour of heated milk.

Other Milk Proteins

In addition to the major protein fractions outlined, milk contains a number of enzymes. The main enzymes present are lipases, which cause rancidity, particularly in homogenised milk, and phosphatase enzymes, which catalyse the hydrolysis of organic phosphates. Measuring the inactivation of alkaline phosphatase is a method of testing the effectiveness of pasteurisation of milk.

Peroxidase enzymes, which catalyse the breakdown of hydrogen peroxide to water and oxygen, are also present. Lactoperoxidase can be activated and use is made of this for milk preservation. Milk also contains protease enzymes, which catalyse the hydrolysis of proteins, and lactalbumin, bovine serum albumin, the immune globulins and lactoferrin, which protect the young calf against infection.

Milk Carbohydrates

Lactose is the major carbohydrate fraction in milk. It is made up of two sugars, glucose and galactose (Figure). The average lactose content of milk varies between 4.7 and 4.9%, though milk from individual cows may vary more. Mastitis reduces lactose secretion.

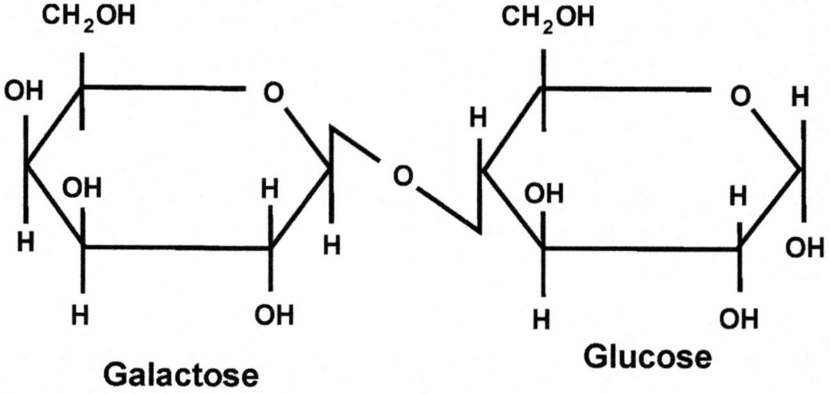

Figure 4. Structure of a lactose molecule.

Lactose is a source of energy for the young calf, and provides 4 calories/g of lactose metabolised. It is less soluble in water than sucrose and is also less sweet. It can be broken down to glucose and galactose by bacteria that have the enzyme β-galactosidase. The glucose and galactose can then be fermented to lactic acid. This occurs when milk goes sour. Under controlled conditions they can also be fermented to other acids to give a desired flavour, such as propionic acid fermentation in Swiss-cheese manufacture.

Lactose is present in milk in molecular solution. In cheese-making lactose remains in the whey fraction. It has been recovered from whey for use in the pharmaceutical industry, where its low solubility in water makes it suitable for coating tablets. It is used to fortify baby-food formula. Lactose can be sprayed on silage to increase the rate of acid development in silage fermentation. It can be converted into ethanol using certain strains of yeast, and the yeast biomass recovered and used as animal feed. However, these processes are expensive and a large throughput is necessary for them to be profitable. For smallholders, whey is best used as a food without any further processing.

Heating milk to above 100°C causes lactose to combine irreversibly with the milk proteins. This reduces the nutritional value of the milk and also turns it brown.

Because lactose is not as soluble in water as sucrose, adding sucrose to milk forces lactose out of solution and it crystallises. This causes sandiness in such products as ice cream.

Special processing is required to crystallise lactose when manufacturing products such as instant skim milk powders. Some people are unable to metabolise lactose and suffer from an allergy as a result.

Pre-treatment of milk with lactase enzyme breaks down the lactose and helps overcome this difficulty. In addition to lactose, milk contains traces of glucose and galactose. Carbohydrates are also present in association with protein. K-casein, which stabilises the casein system, is a carbohydrate-containing protein.

Minor Milk Constituents

In addition to the major constituents discussed above, milk also contains a number of organic and inorganic compounds in small or trace amounts, some of which affect both the processing and nutritional properties of milk.

Milk Salts

Milk salts are mainly chlorides, phosphates and citrates of sodium, calcium and magnesium. Although salts comprise less than 1 % of the milk they influence its rate of coagulation and other functional properties. Some salts are present in true solution. The physical state of other salts is not fully understood. Calcium, magnesium, phosphorous and citrate are distributed between the soluble and colloidal phases (Table 3). Their equilibria are altered by heating, cooling and by a change in pH.

Table 4: Distribution of milk salts between the soluble and colloidal phases.

	Total (mg/100 ml of milk)	Dissolved	Colloidal
Calcium	1320.1	51.8	80.3
Magnesium	10.8	7.9	2.9
Total phosphorus	95.8	36.3	59.6
Citrate	156.6	141.6	15.0

In addition to the major salts, milk also contains trace elements. Some elements come to the milk from feeds, but milking utensils and equipment are important sources of such elements as copper, iron, nickel and zinc.

Milk Vitamins

Milk contains the fat-soluble vitamins A, D, E and K in association with the fat fraction and water-soluble vitamins B complex and C in association with the water phase. Vitamins are unstable and processing can therefore reduce the effective vitamin content of milk.

Milk Protein

General Protein Definition & Chemistry

Proteins are chains of amino acid molecules connected by peptide bonds.

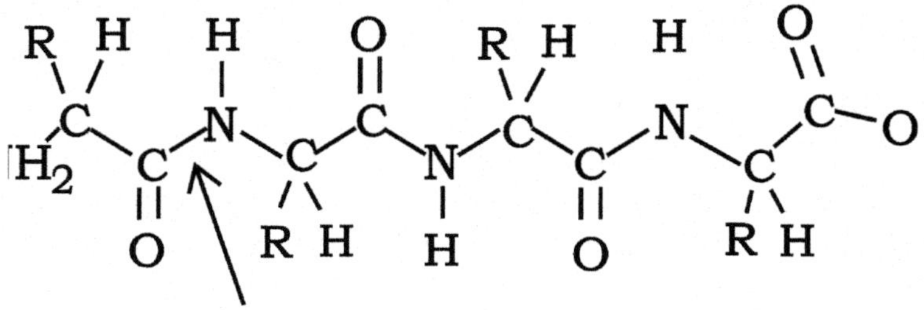

Figure 4: Protein Chain with Peptide Bond

R= amino acid group

There are many types of proteins and each has its own amino acid sequence (typically containing hundreds of amino acids). There are 22 different amino acids that can be combined to form protein chains. There are 9 amino acids that the human body cannot make and must be obtained from the diet. These are called the essential amino acids.

The amino acids within protein chains can bond across the chain and fold to form 3-dimensional structures. Proteins can be relatively straight or form tightly compacted globules or be somewhere in between. The term "denatured" is used when proteins unfold from their native chain or globular shape. Denaturing proteins is beneficial in some instances, such as allowing easy access to the protein chain by enzymes for digestion, or for increasing the ability of the whey proteins to bind water and provide a desirable texture in yogurt production.

Milk Protein Chemistry

Milk contains 3.3% total protein. Milk proteins contain all 9 essential amino acids required by humans. Milk proteins are synthesized in the mammary gland, but 60% of the amino acids used to build the proteins are obtained from the cow's diet. Total milk protein content and amino acid composition varies with cow breed and individual animal genetics. There are 2 major categories of milk protein that are broadly defined by their chemical composition and physical properties. The casein family contains phosphorus and will coagulate or precipitate at pH 4.6. The serum (whey) proteins do not contain phosphorus, and these proteins remain in solution in milk at pH 4.6. The principle of coagulation, or curd formation, at reduced pH is the basis for cheese curd formation. In cow's milk, approximately 82% of milk protein is casein and the remaining 18% is serum, or whey protein.

The casein family of protein consists of several types of caseins (α-s1, α-s2, β, and 6) and each has its own amino acid composition, genetic variations, and functional properties. The caseins are suspended in milk in a complex called a micelle that is discussed below in the physical properties section. The caseins have a relatively random, open structure due to the amino acid composition (high proline content).

The high phosphate content of the casein family allows it to associate with calcium and form calcium phosphate salts. The abundance of phosphate allows milk to contain much more calcium than would be possible if all the calcium were dissolved in solution, thus casein proteins provide a good source of calcium for milk consumers.

The 6-casein is made of a carbohydrate portion attached to the protein chain and is located near the outside surface of the casein micelle. In cheese manufacture, the 6-casein is cleaved between certain amino acids, and this results in a protein fragment that does not contain the amino acid phenylalanine. This fragment is called milk glycomacropeptide and is a unique source of protein for people with phenylketonuria.

The serum (whey) protein family consists of approximately 50% β-lactoglobulin, 20% α-lactalbumin, blood serum albumin, immunoglobulins, lactoferrin, transferrin, and many minor proteins and enzymes. Like the other major milk components, each whey protein has its own characteristic composition and variations. Whey proteins do not contain phosphorus, by definition, but do contain a large amount of sulfur-containing amino acids. These form disulfide bonds within the protein causing the chain to form a compact spherical shape. The disulfide bonds can be broken, leading to loss of compact structure, a process called denaturing. Denaturation is an advantage in yogurt production because it increases the amount of water that the proteins can bind, which improves the texture of yogurt. This principle is also used to create specialized whey protein ingredients with unique functional properties for use in foods. One example is the use of whey proteins to bind water in meat and sausage products.

The functions of many whey proteins are not clearly defined, and they may not have a specific function in milk but may be an artifact of milk synthesis. The function of β-lactoglobulin is thought to be a carrier of vitamin A. It is interesting to note that β-lactoglobulin is not present in human milk. α-Lactalbumin plays a critical role in the synthesis of lactose in the mammary gland. Immunoglobulins play a role in the animal's immune system, but it is unknown if these functions are transferred to humans. Lactoferrin and transferrin play an important role in iron absorption and there is interest in using bovine milk as a commercial source of lactoferrin.

Milk Protein Physical Properties

The caseins in milk form complexes called micelles that are dispersed in the water phase of milk. The casein micelles consist of subunits of the different caseins (α-s1, α-s2 and β) held together by calcium phosphate bridges on the inside, surrounded by a layer of 6-casein which helps to stabilize the micelle in solution.

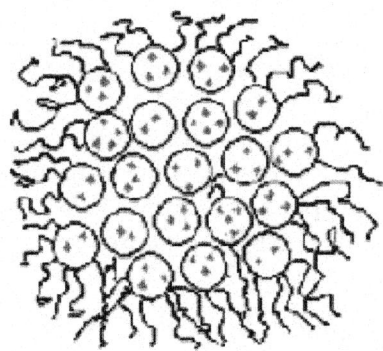

Figure 5: Casein Micelle

Casein micelles are spherical and are 0.04 to 0.3 µm in diameter, much smaller than fat globules which are approximately 1 µm in homogenized milk. The casein micelles are porous structures that allow the water phase to move freely in and out of the micelle. Casein micelles are stable but dynamic structures that do not settle out of solution.

They can be heated to boiling or cooled, and they can be dried and reconstituted without adverse effects. β-casein, along with some calcium phosphate, will migrate in and out of the micelle with changes in temperature, but this does not affect the nutritional properties of the protein and minerals. The whey proteins exist as individual units dissolved in the water phase of milk.

Deterioration of Milk Protein

Proteins can be degraded by enzyme action or by exposure to light. The predominant cause of protein degradation is through enzymes called proteases. Milk proteases come from several sources: the native milk, airborne bacterial contamination, bacteria that are added intentionally for fermentation, or somatic cells present in milk. The action of proteases can be desirable, as in the case of yogurt and cheese manufacture, so, for these processes, bacteria with desirable proteolytic properties are added to the milk. Undesirable degradation (proteolysis) results in milk with off-flavours and poor quality. The most important protease in milk for cheese manufacturing is plasmin because it causes proteolysis during ripening which leads to desirable flavours and texture in cheese.

Two amino acids in milk, methionine and cystine are sensitive to light and may be degraded with exposure to light. This results in an off-flavour in the milk and loss of nutritional quality for these 2 amino acids.

Influence of Heat Treatment on Milk Proteins

The caseins are stable to heat treatment. Typical high temperature short time (HTST) pasteurization conditions will not affect the functional and nutritional properties of the casein proteins. High temperature treatments can cause interactions between casein and whey proteins that affect the functional but not the nutritional properties. For example, at high temperatures, β-lactoglobulin can form a layer over the casein micelle that prevents curd formation in cheese.

The whey proteins are more sensitive to heat than the caseins. HTST pasteurization will not affect the nutritional and functional properties of the whey proteins. Higher heat treatments may cause denaturation of β-lactoglobulin, which is an advantage in the production of some foods (yogurt) and ingredients because of the ability of the proteins to bind more water. Denaturation causes a change in the physical structure of proteins, but generally does not affect the amino acid composition and thus the nutritional properties. Severe heat treatments such as ultra high pasteurization may cause some damage to heat sensitive amino acids and slightly decrease the nutritional content of the milk. The whey protein α-lactalbumin, however, is very heat stable.

Milk Processing

In rural areas, milk may be processed fresh or sour. The choice depends on available equipment, product demand and on the quantities of milk available for processing. In Africa, smallholder milk-processing systems use mostly sour milk. Allowing milk to ferment prior to processing has a number of advantages and processing sour milk will continue to be important in this sector. Where greater volumes of milk can be assembled, processing fresh milk gives more product options, allows greater throughput of milk and, in some instances, greater recovery of milk solids in product.

Because of differences between processing systems, each will be dealt with separately. The section on fresh-milk technology deals with techniques used for processing fresh milk in batches of up to 500 litres. Sour-milk technology is used for processing batches of up to 15 litres of accumulated sour milk. This will be described in the section on sour-milk technology.

Fresh Milk Technology

This section describes the manufacture of skim milk, cream, butter, butter oil, ghee, boiled-curd and pickled cheese varieties and fermented milks from fresh milk. The processing scale envisaged is 100 to 200 litres of milk per day. However, the processes described are suitable for batches of up to 500 litres per day. Most of the equipment described can be fabricated locally. Equipment not available locally, such as a milk separator, has a cost advantage and quickly gives a good financial return in terms of increased efficiency. Hand-operated milk separators are durable and have a long life when properly maintained. Importation of such equipment is, therefore, advantageous.

The procedures given here are very precise. In many rural dairy processing plants, however, monitoring equipment may not be available and, although yields may be maximised by adhering to the prescribed procedures, all these products can be successfully made by approximating temperature, time, pH etc to the best of one's ability. It is particularly important in cheese-making to proceed when the curd is in a suitable condition. Therefore, times given are only approximate and the processor will, with experience, adopt methods suitable to his/her own environment.

Milk Separation

The fat fraction separates from the skim milk when milk is allowed to stand for 30 to 40 minutes. This is known a 'creaming'. The creaming process can be used to remove fat from milk in a more concentrated form. A number of methods are employed to separate cream from milk. An understanding of the creaming process is necessary to maximise the efficiency of the separation process.

Gravity Separation

Fat globules in milk are lighter than the plasma phase, and hence rise to form a cream layer. The rate of rise (V) of the individual fat globule can be estimated using Stokes' Law which defines the rate of settling of spherical particles in a liquid:

$$V = (r^2 (d_1 - d_2)g)/9ç$$

where r = radius of fat globules

d_1 = density of the liquid phase

d_2 = density of the sphere

g = acceleration due to gravity, and

ç = specific viscosity of the liquid phase

Particle r^2: As temperature increases, fat expands and therefore r^2 increases. Since the sedimentation velocity of the particle increases in proportion to the square of the particle diameter, a particle of radius 2 ($r^2 = 4$) will settle four times as fast as a particle of radius 1 ($r^2 = 1$). Thus, heating increases sedimentation velocity.

$d_1 - d_2$: Sedimentation rate increases as the difference between d_1 and d_2 increases. Between 20 and 50°C, milk fat expands faster than the liquid phase on heating. Therefore, the difference between d_1 and d_2 increases with increasing temperature.

g: Acceleration due to gravity is constant. This will be considered when discussing centrifugal separation.

ç: Serum viscosity decreases with increasing temperature. Calculation of the sedimentation velocity of a fat globule reveals that it rises very slowly, As shown in the equation, the velocity of rise is directly proportional to the square of the radius of the globule. Larger globules overtake smaller ones quickly. When a large globule comes into contact with a smaller globule the two join and rise together even faster, primarily because of their greater effective radius. As they rise they come in contact with other globules, forming clusters of considerable size. These clusters rise much faster than individual globules. However, they do not behave strictly in accordance with Stokes' Law because they have an irregular shape and contain some milk serum.

Factors affecting creaming: Cream layer volume is greatest in milk that has high fat content and relatively large fat globules, because such milk contains more large clusters. However, temperature and agitation affect creaming, irrespective of the fat content of the milk. Heating to above 60°C reduces creaming; milk that is heated to above 100°C retains very little creaming ability. Excessive agitation disrupts normal cluster formation, but creaming in cold milk may be increased by mild agitation since such treatment favours larger, loosely packed clusters.

Batch separation by gravity: Cream can be separated from milk by allowing the milk to stand in a setting pan in cool place. There are two main methods.

Shallow pan: Milk, preferably fresh from the cow, is poured into a shallow pan 40 to 60 cm in diameter and about 10 cm deep. The pan should be in a cool place. After 36 hours practically all of the fat capable of rising by this method will have come to the surface, and the cream is skimmed off with a spoon or ladle (Figure). The skim milk usually contains about 0.5 to 0.6% butterfat.

a. Shallow pan   b. Deep setting

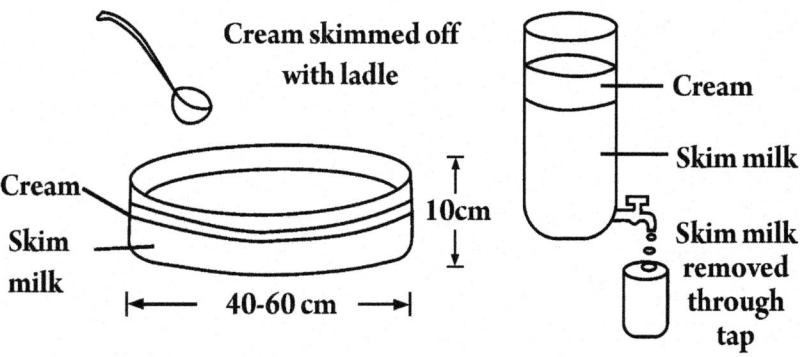

Figure 6. Batch separation of milk by gravity: (a) Shallow pan method, (b) deep-setting method

Deep-setting: Milk, preferably fresh from the cow, is poured into a deep can of small diameter. The can is placed in cold water and kept as cool as possible. After 24 hours the separation is usually as complete as it is possible to secure by this method. The skim milk is removed through a tap at the bottom of the can. Under optimum conditions, the fat content of the skim milk averages about 0.2 or 0.3 %. The pans should be rinsed with water immediately after use, scrubbed with hot water and scalded with boiling water.

Centrifugal Separation

Gravity separation is slow and inefficient. Centrifugal separation is quicker and more efficient, leaving less than 0.1% fat in the separated milk, compared with 0.5–0.6% after gravity separation.

The centrifugal separator was invented in 1897. By the turn of the century it had altered the dairy industry by making centralised dairy processing possible for the first time. It also allowed removal of cream and recovery of the skim milk in a fresh state.

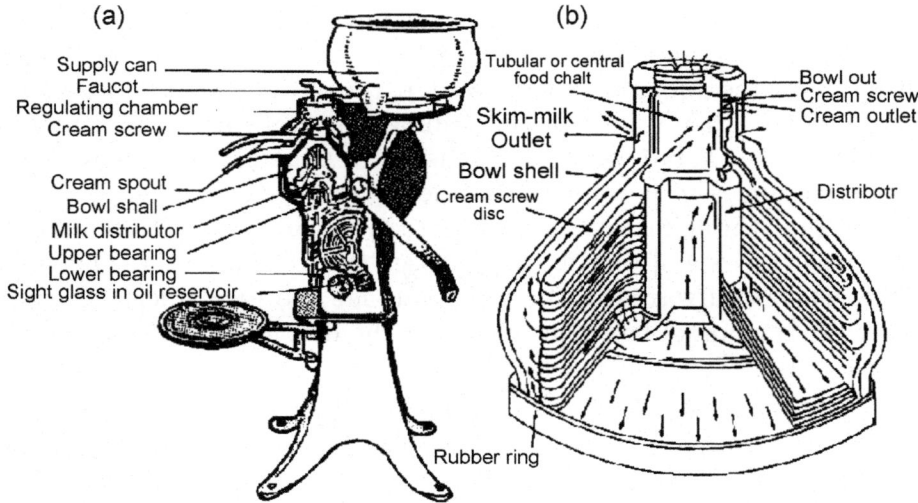

Figure. Cutaway diagrams of (a) hand-operated milk separator and (b) the bowl showing the paths of milk and cream fractions.

The separation of cream from milk in the centrifugal separator is based on the fact that when liquids of different specific gravities revolve around the same centre at the same distance with the same angular velocity, a greater centrifugal force is exerted on the heavier liquid than on the lighter one. Milk can be regarded as two liquids of different specific gravities, the serum and the fat.

Milk enters the rapidly revolving bowl at the top, the middle or the bottom of the bowl (Figure). When the bowl is revolving rapidly the force of gravity is overcome by the centrifugal force, which is 5000 to 10 000 times greater than gravitational force. Every particle in the rotating vessel is subjected to a force which is determined by the distance of the particle from the axis of rotation and its angular velocity.

If we substitute centrifugal acceleration expressed as $r_1\grave{u}^2$ (where r_1 is the radial distance of the particle from the centre of rotation and $\grave{u}^2$ is a measurement of the angular velocity) for acceleration due to gravity (g), we obtain:

$$V = (r^2(d_1 - d_2)\ r_1\grave{u}^2)/9ç$$

Thus, sedimentation rate is affected by $r_1\grave{u}^2$. In gravity separation, the acceleration due to gravity is constant. In centrifugal separation, the centrifugal force acting on the particle can be altered by altering the speed of rotation of the separator bowl.

In separation, milk is introduced into separation channels at the outer edge of the disc stack and flows inwards. On the way through the channels, solid impurities are separated from the milk and thrown back along the undersides of the discs to the periphery of the separator bowl, where they collect in the sediment space. As the milk passes along the full radial width of the discs, the time passage allows even small particles to be separated. The cream, i.e. fat globules, is less dense than the skim milk and therefore settles inwards in the channels towards the axis of rotation and passes to an axial outlet. The skim milk moves outwards to the space outside the disc stack and then through a channel between the top of the disc stack and the conical hood of the separator bowl.

Efficiency of separation is influenced by four factors: the speed of the bowl, residence time in the bowl, the density differential between the fat and liquid phase and the size of the fat globules.

Speed of the separator. Reducing the speed of the separator to 12 rpm less than the recommended speed results in high fat losses, with up to 12% of the fat present remaining in the skim milk.

Residence time in the separator: Overloading the separator reduces the time that the milk spends in the separator and consequently reduces skimming efficiency. However, operating the separator below capacity gives no special advantage—it does not increase the skimming efficiency appreciably but increases the time needed to separate a given quantity of milk.

Effect of temperature: Freshly drawn, uncooled milk is ideal for exhaustive skimming. Such milk is relatively fluid and the fat is still in the form of liquid butterfat. If the temperature of the milk falls below 22°C skimming efficiency is seriously reduced. Milk must therefore be heated to liquify the fat. Heating milk to 50°C gives the optimum skimming efficiency.

Effect of the position of the cream screw: The cream screw regulates the ratio of skim milk to cream. Most separators permit a rather wide range of fat content of cream (18–50%) without adversely affecting skimming efficiency. However, production of cream containing less than 18% or more than 50% fat results in less efficient separation.

Other factors that affect the skimming efficiency are:

- The quality of the milk: Milk in poor physical condition or which is curdy will not separate completely.
- Maintenance of the separator: A separator in poor mechanical condition will not separate milk efficiently.

When separation is complete the separator must be dismantled and cleaned thoroughly.

Hand Separator

In order to understand how centrifugal separation works, we shall follow the course of milk through a separator bowl. As milk flows into a rapidly revolving bowl it is acted upon by both gravity and the centrifugal force generated by rotation. The centrifugal force is 5000 to 10 000 times that of gravity, and the effect of gravity thus becomes negligible. Therefore, milk entering the bowl is thrown to the outer wall of the bowl rather than falling to the bottom. Milk serum has a higher specific gravity than fat and is thrown to the outer part of the bowl while the cream is forced towards the centre of the bowl.

Assembling the Bowl

1. Fit the milk distributor to the central feed shaft.
2. Fit the discs on top of each other on the central shaft.
3. Fit the cream screw disc.
4. Next, fit the rubber ring to the base of the bowl.
5. Put on the bowl shell, ensuring that it fits to the inside of the base.
6. Finally, screw the bowl nut on top.

Now the bowl is assembled and ready for use. The rest of the separator is essentially a set of gears so arranged as to permit the spindle, on which the bowl is carried, to be turned at high speed. The gears are normally enclosed in an oil-filled case. The bowl is usually supported from the bottom and has two bearings; one to support its weight and the second to hold it upright.

The upper bearing is usually fitted inside a steel spring so that it can keep the bowl upright even if the frame of the machine is not exactly level. The assembled bowl is lowered into the receptacle, making sure that the head of the spindle fits correctly into the hollow of the central feed shaft.

Operation

1. When the bowl is set, fit the skim milk spout and the cream spout.
2. Fit the regulating chamber on top of the bowl.
3. Put the float in the regulating chamber.
4. Put the supply can in position, making sure that the tap is directly above and at the centre of the float.
5. Pour warm (body temperature) water into the supply can.
6. Turn the crank handle, increasing speed slowly until the operating speed is reached: This will be indicated on the handle or in the manufacturer's manual of operation. The bell on the crank handle will stop ringing when the correct speed is reached.
7. Open the tap and allow the warm water to flow into the bowl. This rinses and heats the bowl and allows a smooth flow of milk and increases separation efficiency.
8. Next, put warm milk (37 – 40°C) into the supply can. Repeat steps 6 and 7 above and collect the skim milk and cream separately.
9. When all the milk is used up and the flow of cream stops, pour about 3 litres of the separated milk into the supply can to recover residual cream trapped between the discs.
10. Continue turning the crank handle and flush the separator with warm water.

Cleaning the separator: Many of the impurities in the milk collect as slime on the wall of the separator bowl. This slime contains remnants of milk, skim milk and cream, all of which will decompose and ferment unless removed promptly.

If not washed and freed from all impurities the separator bowl becomes a source of microbial contamination. Skimming efficiency is also reduced when the separator bowl and discs are dirty. Milk deposits on the separator can cause corrosion.

Washing the separator: After flushing the separator with warm skim milk, the bowl should be flushed with clean water until the discharge from the skim milk spout is clean. This removes any residual milk solids and makes subsequent cleaning of the bowl easier. The bowl should then be dismantled. Wash all. parts of the bowl, bowl cover, discharge spouts, float supply tank and buckets with a brush, hot water and detergent. Rinse with scalding water. Allow the parts to drain in a clean place protected from dust and flies. This process should be followed after each separation.

Cream Screw Adjustment

The cream screw should be adjusted so that the fat content of the cream is about 33%. Producing excessively thin cream reduces the amount of separated milk available for other uses and increases the volume of cream to be handled. Low-fat cream is also more difficult to churn efficiently.

Cream containing more than 45% fat clogs the separator and causes excessive loss of fat in skim milk. Cream of abnormally high fat content also gives butter a greasy body due to lack of milk SNF. When adjusting the cream screw it is important to remember that it is very sensitive; a quarter turn of the screw is sufficient to change the percentage fat in the cream appreciably.

The fat content of whole milk influences the fat content of cream and this must be considered when adjusting the cream screw. For example, if the cream screw is set to separate milk at a ratio of 85 parts of separated milk to 15 parts of cream then, with all other conditions constant and assuming efficient separation, milk of 3% fat produces cream of 20% fat whereas milk of 4.5% fat produces cream of 30% fat. The fat content of the cream can be calculated using the following equation:

$$Fc = (Wm \times Fm)/Wc$$

Wm = weight of milk	Fm = fat content of milk
Wc = weight of cream	Fc = fat content of cream
In the first example,	Fc = (100 × 3)/15 = 20
In the second example,	Fc = (100 × 4.5)/15 = 30

Therefore the setting of the cream screw depends on the fat content of the milk being separated. The milk should be mixed thoroughly prior to separation to ensure even distribution of cream in the milk.

Separator Maintenance

- The gears must be well lubricated. Follow the directions of the manufacturer.
- The level of the lubricant must be kept constant; observe the oil level through the sight glass.
- The bowl must be perfectly balanced.
- The bowl should be cleaned thoroughly immediately after use to ensure proper functioning of the separator and for hygiene.

Calculations

Once milk passes through a separator it is recovered in two fractions, the high-fat cream fraction and the low-fat skim milk. Assuming negligible loss of fat in the separator, the amount of fat entering the separator with the whole milk will be collected at the other side of the separator in either the cream or the skim milk. Therefore, if we separate 200 kg of milk containing 4.5% butterfat, what weight of cream containing 30% butterfat can we expect?

Let Wm = weight of milk

Fm = fat content of the milk

Wc = weight of cream

Fc = fat content of the cream

Ws = weight of skim milk

Assuming that all of the fat present in the milk is recovered in the cream, then:

$Wm \times Fm = Wc \times Fc$

and $Wm - Wc = Ws$

and $Wm - Ws = Wc$

Since $Wm \times Fm = Wc \times Fc$

$(Wm \times Fm)/Fc = Wc$

Therefore $Ws = Wm - (Wm \times Fm)/Fc = Wc$

In this case: $Ws = 200 - (200 \times 4.5)/30 = 200-30 = 170$ kg

Since $Wc = Wm - Ws$

$Wc = 200 - 170 = 30$ kg

Percentage yield of skim milk:

$=Ws \times 100)/Wm = (170 \times 100)/200 = 85\%$

Percentage cream ($\%Wc$) = $\%Wm - \%Ws = 100 - 85 = 15\%$

If in practice we obtain only 28 kg of cream containing 30% butterfat, then (2×0.30) kg or 0.6 kg of butterfat has not been recovered in the cream. Since it is assumed that there are no significant losses of fat in the cream separator, the fat not recovered in the cream is lost in the skim milk.

Since 28 kg of cream was produced, and

$Ws = Wm - Wc$

then $Ws = 200 - 28 = 172$ kg

Thus there is 0.6 kg of fat in 172 kg of skim milk. The fat percentage of the skim milk is therefore:

$(0.6 \times 100)/172 = 0.35\%$*

* The skim milk contains 0.35% fat, which may be incorporated in cottage cheese. If the skim milk is consumed, no nutritional loss occurs, but a financial loss is incurred since the fat is more valuable if sold as butter than as cottage cheese or if it is consumed directly. The percentage of fat in milk and in cream influences Wc and Ws where the fat is recovered in the cream.

If $Fm = 3\ \%$

       $Fc = 30\%$

       $Wm = 100$

Then $Wc = Wm \times Fm/Fc$

$Wc = 100 \times 3/30 = 10$ kg

$Ws = Wm - Wc = 100 - 10 = 90$ kg

whereas if $Fm = 4\%$

          $Fc = 30$

          $Wc = 100$

Then $Wc = 100 \times 4/30 = 13.3$ kg

$Ws = 100 - 13.3 = 86.6$ kg

Standardisation of Milk and Cream

If fine adjustment of the fat content of cream is required, or if the fat content of whole milk must be reduced to a given level, skim milk must be added. This process is known as standardisation.

The usual method of making standardisation calculations is the Pearson's Square technique. To make this calculation, draw a square and write the desired fat percentage in the standardised product at its centre and write the fat percentage of the materials to be mixed on the upper and lower left-hand corners. Subtract diagonally across the square the smaller from the larger figure and place the remainders on the diagonally opposite corners. The figures on the right-hand corners indicate the ratio in which the materials should be mixed to obtain the desired fat percentage.

The value on the top right-hand corner relates to the material on the top left-hand corner and the figure on the bottom right relates to the material at the bottom left corner.

Example 1

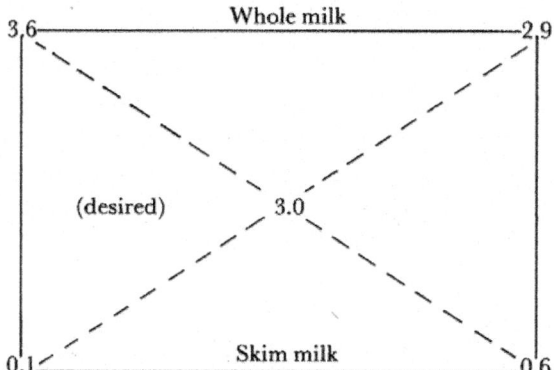

In this example, the fat content of whole milk is to be reduced to 3.0%, using skim milk produced from some of the whole milk. Using Pearson's Square, it can be seen that for every 2.9 litres of whole milk, 0.6 litres of skim milk must be added.

Example 2

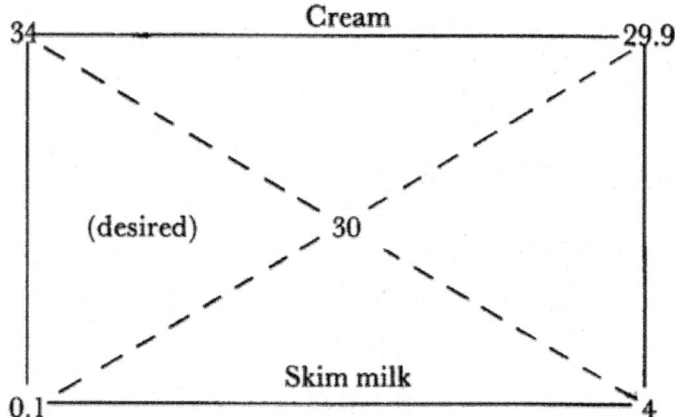

How much skim milk containing 0.1 % fat is needed to reduce the percentage fat in 200 kg of cream from 34% to 30%?

If 29.9 parts of cream require 4 parts of skim milk, 200 parts of cream require x parts of skim milk.

Weight of skim milk needed = x = (200 × 4)/29.9 = 26.75 kg

Example 3

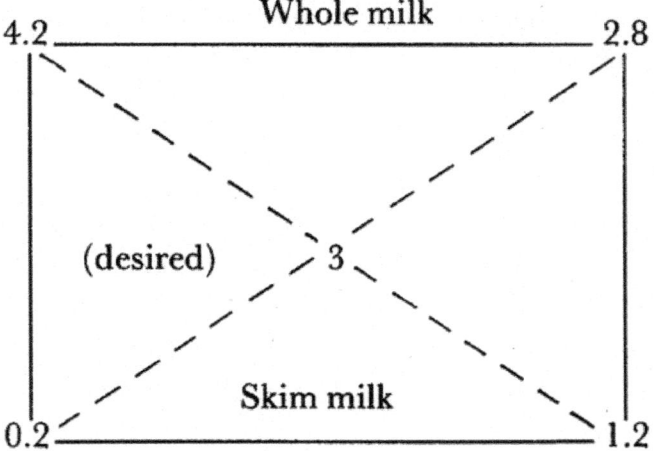

The fat content of 300 kg of whole milk must be reduced from 4.2% to 3% using skim milk containing 0.2% fat.

Every 4.0 kg of the mixture will contain 2.8 kg of whole milk and 1.2 kg of skim milk.

If 2.8 kg of whole milk requires 1.2 kg skim milk, 300 kg of whole milk requires (1.2 × 300)/2.8 = 128.6 kg of skim milk

Thus, 128.6 kg of skim milk (0.2% fat) must be added to 300 kg of whole milk (4.2% fat) to give 428.6 kg of milk containing 3% fat.

Example 4

The fat content of milk must be reduced from 4.5 to 3% prior to sale as liquid milk but skim milk for standardisation is not available.

In this case, we must calculate (a) what proportion of the milk must be separated to provide enough skim milk to standardise the remaining whole milk and (b) the expected yield of cream.

Assume that the fat content of 100 kg of milk containing 4.5% milk fat must be reduced to 3%. The amount of cream to be removed can be calculated as follows:

Let M = weight of milk to be standardised—in this example, 100 kg. Therefore M = 100

Fm = fat content of the original milk = 4.5

C = weight of cream

Fc = fat content of the cream = 35

SM = weight of standardised milk

Fsm = fat content of the standardised milk = 3.0

Since the milk is separated into cream and standardised milk

SM + C = M

(1) or SM + C = 100

There are no fat losses; therefore the weight of fat in the original milk will be equal to the weight of fat in the standardised milk and cream.

(Weight of fat in a product is the weight of product × % fat/100)

Therefore (SM × Fsm)/100 + (C × Fc)/100 = (M × Fm)/100

or (3 × SM)/100 + (35 × C)/100 = (100 × 4.5)/100

(2) or 0.03SM + 0.35C = 4.5

Equations (1) and (2) give two equations with two unknowns, so they can be solved as follows:

(1) SM + C = 100

(3) or 0.03SM + 0.03C = 3

Subtracting (3) from (2)

0.32 C = 1.5

C = 4.6875

= 4.7 corrected to one decimal place

The weight of cream is thus 4.7 kg.

Therefore, the weight of standardised milk is 95.3 kg.

Answer check

The original milk contained 4.5 kg of fat.

The cream contains (4.7 × 35)/100 =1.645 kg of fat

Therefore 4.5 − 1.645 = 2.855 kg of fat in the standardised milk.

The fat percentage of the standardised milk is

(2.855 × 100)/95.3 = 3%

The calculation can also be made using Pearson's Square. This is essentially a reverse standardisation, i.e. "how much cream containing 35% fat and milk containing 3% fat should be mixed to get milk containing 4.5% fat?" is mathematically the same as "how much cream containing 35% fat must be removed from milk containing 4.5% fat to standardise the milk to 3% fat content?"

1. Place the fat content of whole milk in the centre.

2. Place the fat content of cream on the top left-hand corner.

3. Place the desired fat content of the standardised milk on the bottom left-hand corner.

4. For every 32 parts of whole milk, there are 1.5 parts of cream to be removed and 30.5 parts of standardised milk.

Therefore $Wc = (1.5)/32 \times 100 = 4.6875 = 4.7$

$Wsm = Wm - Wc = 95.3$

The Wsm and fat to be removed can be calculated in a number of ways.

Whatever method is used to calculate the amount of cream to be removed, it is then necessary to calculate the amount of milk to be separated to achieve the desired reduction in fat content.

$Wm \times Fm = Wc \times Fc$

Therefore $Wm \times 4.5 = 4.7 \times 35$

and $Wm = (4.7 \times 35)/4.5 = 36.5$

Therefore, 36.5 kg of milk are separated and the skim milk is then combined with the remaining whole milk. Standardisation such as this can be used to increase income from milk production as follows:

Assume liquid milk price of 70 cents/kg

Assume butter price of EB* 10/kg

Income from 100 kg of milk = EB 70

Income from 95.3 kg of milk = 66.71

Fat removed = $Wc \times Fc = 4.7 \times 0.35 = 1.645$

Expected butter yield = 1.9 kg

Income from butter = EB 19

Total income = EB 85.76

Margin = EB 15.76/ 100 kg of milk

*EB = Ethiopian birr (US$ 1 = EB 2.07)

Butter-making with Fresh Milk or Cream

Butterfat can be recovered from milk and converted to a number of products, the most common of which is butter. Butter is an emulsion of water in oil and has the following approximate composition:

Fat	80%
Moisture	16%
Salt	2%
Milk SNF	2%

In good butter the moisture is evenly dispersed throughout the butter in tiny droplets. In most dairying countries legislation defines the composition of butter; and butter makers conform to these standards insofar as is possible.

Butter can be made from either whole milk or cream. However, it is more efficient to make butter from cream than from whole milk.

Butter-making Theory

To make butter, milk or cream is agitated vigorously at a temperature at which the milk fat is partly sold and partly liquid. Churning efficiency is measured in terms of the time required to produce butter granules and by the loss of fat in the buttermilk. Efficiency is influenced markedly by churning temperature and by the acidity of the milk or cream.

In churning, cream is agitated in a partly filled chamber. This incorporates a large amount of air into the cream as bubbles. The resultant whipped cream occupies a larger volume than the original cream. As agitation continues the whipped cream becomes coarser. Eventually the fat forms semi-solid butter granules, which rapidly increase in size and separate sharply from the liquid buttermilk. The remainder of the butter-making process consists of removing the buttermilk, kneading the butter granules into a coherent mass and adjusting the water and salt contents to the levels desired.

Theory of the Mechanism of Churning

In considering the mechanism of churning the following factors must be taken into account:

- The function of air;
- The release of stabilising material from the fat globule surface into the buttermilk;

- The differences in structure between butter and cream; and
- The temperature dependence of the process.

Air is thought to be necessary for the process, but some workers have demonstrated that milk or cream can be churned in the absence of air, although it takes longer.

About one half of the stabilising material is liberated into the buttermilk during churning. It is thought that during churning the fat globule membrane substance spreads out over the surface of the air bubbles, partly denuding the globules of their protective layer, and that a liquid portion of the fat exudes from the globule and partly or entirely covers the globule, rendering it hydrophobic.

In this condition the globules tend to stick to the air bubbles. Free fat destabilises the foam, causing it to collapse. The partly destabilised globules clinging to the air bubbles thus collect in clusters cemented together by free fat. These clusters appear as butter grains.

Churning Cream

Cream prepared by gravitational or mechanical separation can be used. Good butter can be made in any type of churn provided it is clean and in good repair.

Churn Preparation

The churn is prepared by rinsing with cold water, scrubbing with salt and rinsing again with cold water. Alternatively, it can be scalded with water at 80°C. After the butter has been removed, the churn should be washed well with warm water, scalded with boiling water and left to air. When not in use wooden churns should be soaked occasionally with water. A new churn should first be washed with tepid water, scrubbed with salt and then washed with hot water until the water comes away clear. A hot solution of salt should then be allowed to stand in the churn for a short time. After rinsing again with hot water the churn should be left to air for at least one day before being used.

Churning Temperature

The temperature of the cream during churning is of great importance. If too cool, butter formation is delayed and the grain is small and difficult to handle. If the temperature is too high, the yield of butter will be low, because a large proportion of the fat will remain in the buttermilk, and the butter will be spongy and of poor quality. Cream should be churned at 10 –12°C in the hot season and at 14 –17°C in the cold season. The temperature may be raised by standing the vessel containing the cream in hot water, or may be lowered by standing the vessel in cold spring water for a few hours before the cream is churned. The churning temperature may also be adjusted by the water used to dilute the cream. In the hot season, the coldest water available should be used, preferably water that has been stored in a refrigerator.

The amount of cream to be churned should not exceed one half the volumetric capacity of the churn. An airtight churn should be ventilated frequently during the first 10 minutes of churning to release gases driven out of solution by the agitation. If butter is slow in forming, adding a little water which is warmer than the churning temperature, but never over 25°C, usually causes it to form more quickly. When the butter appears like wet maize meal, water (1 litre per 4 litres of cream) at 2°C below the churning temperature should be added. It may be necessary to add water a second time to maintain butter grains of the required size. Churning should cease when the butter grains are as large as small wheat grains.

Washing the Butter

When the desired grain size is obtained, the buttermilk is drained off and the butter washed several times in the churn. Each washing is done by adding only as much water as is needed to float the butter and then turning the churn a few times. The water is then drained off: As a general rule two washings will suffice but in very hot weather three may be necessary before the water comes away clear. In the hot season the coldest water available should be used for washing, and in the cold season water about 2 to 3°C colder than the churning temperature should be used.

Salting, Working and Packing the Butter

Equipment for working may consist of a butter worker or a tub or keeler. Good-quality spatulas are important, and a sieve and scoop facilitate the removal of butter from the churn. This equipment must be clean. The butter is spread on the worker, which has been soaked previously with water of the same temperature as the washing water. If salted butter is required, the butter should be salted before working at a rate of 16 g salt/kg or according to taste. The salt used should be dry and evenly ground and of the best quality available.

The butter is then either rolled out 8 to 10 times or ridged with the spatulas to remove excess moisture. If the butter is to be heavily salted, it must be worked more in proportion to the amount of salt used, as uneven distribution of the salt causes uneven colour. The butter should be worked until it seems dry and solid, but it must not be worked too much or it will become greasy and streaky.

The butter is then weighed and packed for storage. It should be packed in polythene-lined wooden or cardboard cartons and stored in a cool, dry place. The butter should be firm and of uniform colour.

Washing the Churn and Butter-making Equipment After Use

The churn and butter-making equipment should be washed as soon as possible, preferably while the wood is still damp.

Churn: Wash the inside of the churn thoroughly with hot water. Invert the churn with the lid on in order to clean the ventilator; this should be pressed a few times with the back of a scrubbing brush to allow water to pass through.

Remove the rubber band from the lid and scrub the groove. Scald the inside of the churn with boiling water. This step is very important. Invert and leave to air. Dry the outside and treat steel parts with vaseline to prevent rusting. The rubber band should not be placed in boiling water; dipping in warm water is sufficient.

Butter worker/keeler: Place the sieve, scoop and spades on the butter worker or keeler. Pour hot water over all of them and scrub well to remove all traces of grease. Scald with boiling water and leave to air. Treat the steel part of the butter worker with vaseline to prevent rusting.

Storage of Butter

Surplus good-quality butter can be stored, but should contain more salt than usual—at least 30 g/kg. Low moisture content is desirable. The butter must be packed in clean containers, such as seasoned boxes or glazed crocks, and stored in a cold room or in a cold, airy place. If a box is used, it should be lined with good-quality polythene. The container should be filled to capacity from one churning. The more firmly butter is packed, the better; it may be covered with a layer of salt, but this is not essential. The container should be securely covered with a lid or a sheet of strong paper.

Overrun and Produce in Butter-making

An enterprise engaged in butter-making must be able to measure the efficiency of the process, i.e. by measuring the yield of butter from the butterfat purchased.

First, the theoretical yield of butter has to be estimated. Butter contains an average of 80% butterfat. Thus, for every 80 kg of butterfat purchased 100 kg of butter should be produced, or for every 100 kg of butterfat purchased 125 kg of butter should be produced.

The difference between the number of kilograms of butterfat churned and the number of kilograms of butter made is known as the overrun. This difference is due to the fact that butter contains non-fatty constituents such as moisture, salt, curd and small amounts of lactic acid and ash in addition to butterfat.

The overrun is financially important to the dairy industry and constitutes the margin between the purchase price of butterfat and the sale price of butter. The dairy unit depends largely on overrun to cover manufacturing costs and to defray expenses incurred in the purchase of milk.

As stated above, the maximum legitimate overrun is 25%. In commercial operation, however, it is not possible to establish the degree of accuracy that is assumed in the calculation of theoretical overrun, and the actual overrun shows the difference between the amount of butter churned out and the amount of butterfat bought.

Overrun is affected by:

· Accuracy of weighing milk received.

· Accuracy of sampling and testing milk for fat.

Produce

Another method for estimating the efficiency of a process is to measure the number of litres of milk required per kilogram of butter produced.

For example, how many litres of milk containing 4% butterfat are required to make 1 kg of butter?

In 1 kg of butter there is 0.80 kg of butterfat.

In the milk we have 4 kg fat/ 100 kg or per 100 litres/1.032.

Therefore we have:

1 kg fat in 100/(1.032 × 4) = 24.22 litres

or 0.8 kg. fat in 19.38 litres

Therefore 19.38 litres of milk containing 4% fat will be required to make 1 kg of butter. Thus the efficiency of operation can also be checked by calculating output. The fat content of the whole milk, skim milk and buttermilk should be checked daily. The moisture content of the butter should be checked for each batch. The accuracy of weighing scales and other measuring devices should be checked regularly.

Butter Quality

Butter quality can be discussed under two main headings:

- Compositional quality
- Organoleptic quality.

The compositional quality of butter can be further divided into two subsections:

- Chemical composition
- Bacteriological composition.

Compositional Quality

The chemical composition of butter is determined at the processing stage when the salt, moisture, curd and fat contents of the product are regulated. Once these parameters have been set there is little one can do to change them. The microbiological quality of butter is also determined during the production and processing stages.

Chemical composition affects butter yield, while butter of poor microbiological quality will deteriorate rapidly and become unacceptable to consumers. The butter may also contain pathogens. Cleanliness at all stages of production is, therefore, essential.

Organoleptic Quality

The organoleptic quality of butter can be described as the customer's reaction to its colour, texture and flavour. It has been said that the consumer tastes with his or her eyes, and it is true that a person's initial impression of a food will often determine whether or not he or she will buy it.

It is important, therefore, to produce butter that has an even colour, clean flavour and close texture. It is also important that it be free from defects such as loose moisture. It should be packed attractively, both to attract customer attention and to retain its quality.

Butter produced carelessly and without the use of preservatives has a very short shelf life. Preservation of butter quality can assist the smallholder in two ways:

- The less perishable the product the longer the smallholder can retain it to obtain a good price.

• He or she can store the surplus made during the production season for consumption during the season in which he or she cannot produce butter.

The first step the producer can take to ensure a high-quality product is to make it in a clean, hygienic manner.

This results in fewer spoilage organisms being present in the butter. Another step is to take care in the handling and storage of the butter.

The use of permitted preservatives is by far the most effective means of maintaining butter quality when used in conjunction with the above precautions. Salt—sodium chloride—is an excellent preservative, and salting butter to 3% extends its storage life: salted butter can be stored for up to 4 months without significant deterioration. A salt concentration in excess of 3% gives little advantage and can adversely affect the flavour of the butter.

Aside from the influence of salt on the flavour and keeping quality of the butter, adding salt is of economic importance as it increases overrun.

Adding salt to butter disturbs the equilibrium of the emulsion (the butter). This, in turn, changes the character of the body and alters its colour. Unless the butter is subjected to sufficient working to regain the original equilibrium of the emulsion, it will tend to have a coarse, leaky body and uneven colour.

Salt is added to butter most commonly using the dry-salting method, in which dry salt is sprinkled evenly over the butter and worked in.

Butter must be adequately worked if it is to be stored for a long time. First, working distributes the salt uniformly in the moisture and this helps inhibit microbial growth. Secondly, it distributes the salt solution into many tiny droplets rather than fewer large ones. For a given level of microbial contamination, the microbes will be more isolated in small droplets and will have less of the butter's nutrients available to them for growth. After salting, the butter should be stored in a clean container, and the container sealed. It should then be stored in a cool, dark place.

Ghee, Butter Oil and Dry Butterfat

These products are almost entirely butterfat and contain practically no water or milk SNF. Ghee is made in eastern tropical countries, usually from buffalo milk. An identical product called *samn* is made in Sudan. Much of the typical flavour comes from the burned milk SNF remaining in the product. Butter oil or anhydrous milk fat is a refined product made by centrifuging melted butter or by separating milk fat from high-fat cream.

Ghee is a more convenient product than butter in the tropics because it keeps better under warm conditions. It has low moisture and milk SNF contents, which inhibits bacterial growth.

Milk or cream is churned as described in the sections dealing with churning of whole milk or cream. When enough butter has been accumulated it is placed in an iron pan and the water evaporated at a constant rate of boiling. Overheating must be avoided as it burns the curd and impairs the flavour. Eventually a scum forms on the surface: this can be removed using a perforated ladle. When all the moisture has evaporated the casein begins to char, indicating that the process is complete. The ghee can then be poured into an earthenware jar for storage.

A considerable amount of moisture and milk SNF can be removed prior to boiling by melting the butter in hot water (80°C) and separating the fat layer. The fat can be separated either by gravity or using a hand separator. The fat phase yields a product containing 1.5% moisture and little fat is lost in the aqueous phase.

Alternatively, the mixture can be allowed to settle in a vessel similar to that used in the deep-setting method for separating whole milk. Once the fat has solidified the aqueous phase is drained. The fat is then removed and heated to evaporate residual moisture. Products made using these methods exhibited excellent keeping qualities over a 5-month test period.

Cheese-making

Cheese is a concentrate of the milk constituents, mainly fat, casein and insoluble salts, together with water in which small amounts of soluble salts, lactose and albumin are found. To retain these constituents in concentrated form, milk is coagulated by direct acidification, by lactic acid produced by bacteria, by adding rennet, or a combination of acidification and addition of rennet.

Rennet Coagulation Theory

Rennet, a proteolytic enzyme extracted from the abomasum of suckling calves, was traditionally used for coagulating milk. Originally, the abomasum was itself immersed in milk. The extraction of rennet that could be stored as a liquid was the first step towards refining this procedure.

This was followed by purification and concentration of the enzyme. The purified enzyme was originally called rennin, and is now called chymosin.

On weaning, the chymosin of the suckling calf is replaced by bovine pepsin. With the decrease in the practice of slaughtering calves, chymosin became scarce, resulting in a search for chymosin substitutes. Rennet is a general term currently used to describe a variety of enzymes of animal, plant or microbial origin used to coagulate milk in cheese-making. Rennet transforms liquid milk into a gel. While the process is not fully understood, rennet coagulation is thought to take place in two distinct phases, the first of which is regarded as being enzymatic, the second non-enzymatic. The first, or primary phase, can be illustrated as:

$$\text{Casein} \xrightarrow[\text{rennet}]{\text{water}} \text{para casein} + \text{glycomacropeptide}$$

Since k-casein stabilises the other caseins and its hydrolysis leads to the coagulation of the casein fraction, the primary phase can also be expressed as:

$$\text{8-casein} \xrightarrow[\text{rennet}]{\text{water}} \underset{\text{(insoluble)}}{\text{para 8-casein}} + \underset{\text{(soluble)}}{\text{glycomacropeptide}}$$

The effect of milk coagulants on the other caseins is thought to be negligible at this stage.

The second, or secondary, phase is the non-enzymatic precipitation of para casein by calcium ions. Para casein, in association with the calcium ions, is thought to produce a lattice structure throughout the milk. This traps the fat and whey is gradually exuded. The coagulum then contracts, a process known as syneresis. This is accelerated by increasing the temperature and reducing pH to as low as pH 4.6.

Rennet also has a tertiary action on milk proteins. This occurs during cheese ripening, during which rennet hydrolyses milk proteins. If the desired hydrolysis is not obtained, the cheese becomes bitter. While a wide variety of proteolytic enzymes coagulate milk, the tertiary action of many of these on milk proteins causes undesirable flavours in cheese, which limits the range of coagulants that can be used.

Cheese Varieties

Many cheese varieties are manufactured around the world but they are all broadly classified by hardness (i.e. very hard, hard, semi-soft and soft) according to their moisture content. Cheese is usually made from cows milk, although several varieties are made from the milk of goats, sheep or horses. Flow diagrams for the manufacture of the varieties discussed.

Queso Blanco (White Cheese)

Queso blanco is a Latin-American fresh, white cheese. It is usually made from milk containing 3% fat, using an organic acid, without starter or rennet.

Procedure

1. Take fresh whole milk and determine its fat content. If the fat content is higher than 3%, standardise using skim milk.
2. Transfer the standardised milk to a cheese vat, preferably a double-jacketed standard cheese vat, and heat to 82°C.
3. While the milk is being heated measure out lemon juice of pH about 2.5 in a measuring jar. About 3 ml of lemon juice should be added per 100 ml of milk.
4. Dilute the lemon juice with an equal amount of clean, fresh water.
5. When the milk temperature reaches 82°C, add the diluted lemon juice carefully and uniformly while stirring. For even distribution of the juice, add in three separate amounts.
6. The curd precipitates almost immediately. Continue to stir for 3 minutes after adding the juice, then allow the curd to settle for 15 minutes
7. Drain the whey through a metal sieve or cheese cloth.
8. While draining the whey, stir the curd to prevent excess matting.
9. Distribute a total of about 3.5 to 5 kg of salt to 100 kg curd, in three applications.
10. Prepare a cylindrical or square hoop by lining with cheese cloth and scoop the salted curds into it.
11. Press the curd overnight at room temperature.
12. Remove the pressed cheese and cut into blocks of 0.5 or 1 kg.

Queso blanco is made without starter or rennet. A variety of acidulants can be used for its manufacture. Heating the milk to 82°C pasteurises the milk and denatures the whey proteins, so that they are recovered with the curd.

This increases cheese yield. The cheese has good keeping quality and is thus suitable for manufacture in rural areas.

Halloumi

Halloumi is the curd, formed by coagulating whole milk using rennet or similar enzymes, from which part of the moisture (whey) has been removed by cutting (bleeding), warming and pressing.

Procedure

1. Heat the milk to 32–35°C.
2. Add rennet or a similar enzyme according to the manufacturer's directions, while stirring the milk.
3. Hold the milk at 32–35°C until the curd sets.
4. Check for setting of the curd by applying pressure to the edge of the milk where it comes in contact with the vat, using a spatula or a knife with a round tip. If the curd is set it comes away clean from the wall of the vat.
5. After coagulation, the curd is cut into 3–5 mm cubes using vertical and horizontal knives.
6. Hold the curd in whey for about 20 minutes, stirring gently and continuously, and then allow it to settle.
7. Drain the whey and scoop out the curd into a hoop lined with cheese cloth. Press the curd.
8. While the curd is in the press, heat the whey to about 80–90°C. This precipitates the whey proteins, which can then be removed and pressed to make a whey cheese.
9. Take out the pressed curd, cut it into pieces of 10 × 10 × 3 cm and heat at about 80°C in hot whey. Continue heating until the pieces of curd float on the surface of the whey and become soft and elastic.
10. Remove the pieces of curd when still warm and either press in the hands, folded or unfolded, and rub in a little dry salt mixed with dried leaves of *Mentha viridis* (spearmint).
11. When the pieces are cold, put them in containers filled with cool, boiled whey brine and store in a cool place to ripen for about 30 days.

12. After ripening put in an airtight container and store in a refrigerator at less than 12°C. The cheese will keep for several months under these conditions. Halloumi cheese is best after 40 days but can also be consumed just after manufacture.

Note: 15% salt concentration in whey brine is normally used.

Expected yield: 1 kg of cheese from 9 kg of milk (11 %).

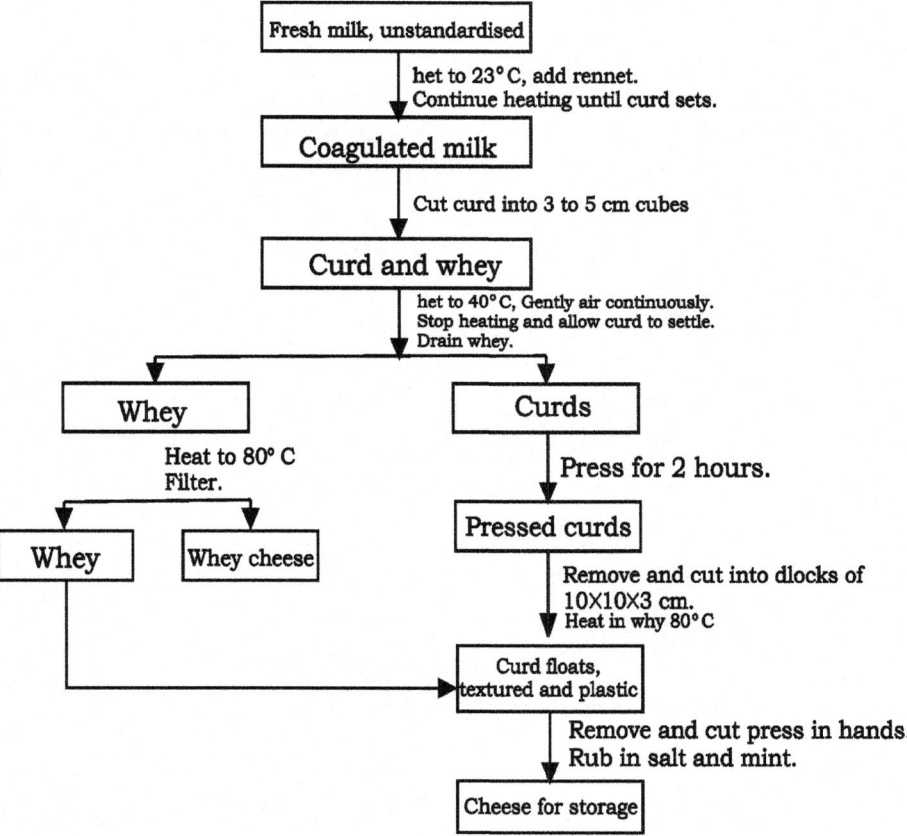

Figure 1. Manufacturing steps for Halloumi cheese.

Domiati–Gybna Beyda

Known as Domiati in Egypt and Gybna beyda in Sudan, this is a hard, white cheese.

Procedure

1. Heat fresh milk to 35°C and add enough salt to give 7 to 10% salt solution in the milk.

2. Add enough rennet to coagulate the milk in 4 to 6 hours.

3. Once set, transfer the coagulum to wooden moulds lined with muslin.

4. Allow the whey to drain overnight.

5. On the following day, pack the cheese in tins and fill the interspaces with whey.

6. Seal the tins by soldering.

Notes:

1 and 2. In some areas rennet is added before salting. In this procedure, salt is not added until a coagulum has formed. If salt is added before rennet it is not advisable to add more rennet to shorten the coagulation time, as this reduces the quality of the cheese.

6. Whey expulsion continues during storage and the cheese hardens.

Expected yield: 1 kg of cheese from 7 kg of milk (15%).

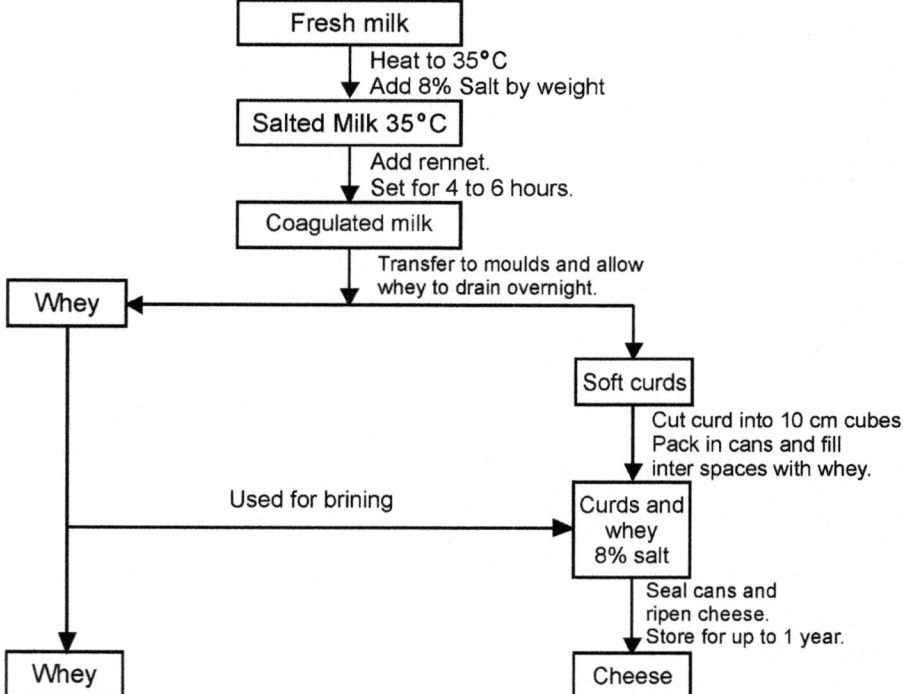

Figure 2: Manufacturing steps for Domiati/Gybna beyda cheese.

Feta

This is a brine-pickled cheese. It can be made from milk of cows, sheep or goats. Feta can be made without starter and can also be made from standardised milk. The procedure described here is for the manufacture of a feta-type cheese without starter or additives.

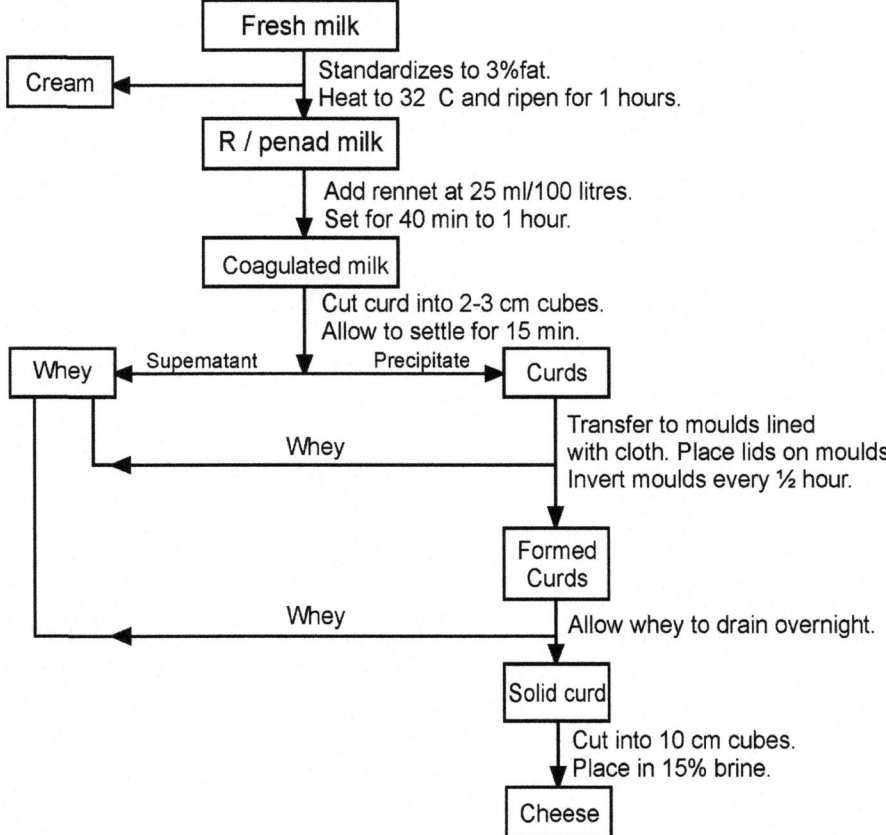

Figure 3. Manufacturing steps for Feta cheese.

Procedure

- Standardise the milk to 3% fat, heat to about 32°C and allow to ripen for one hour before adding rennet.

- Add commercial rennet at the rate of 25 ml/ 100 litres of milk. Leave the milk until a firm clot has formed—this usually takes 40 to 50 minutes.

- Cut the curd into 2-to 3-cm cubes to facilitate whey drainage. Allow 15 minutes for the whey to separate. Stir intermittently during this time.

- Allow the curds to settle and decant the supernatant whey.
- Transfer the curds and some whey to cheese moulds lined with muslin. Place the lid on the mould and invert at half-hourly intervals in the first few hours to facilitate whey drainage.
- Allow the curd to settle overnight.
- On the following day, cut the curd mass into blocks of suitable size and sprinkle them with salt.
- Place the salted blocks in a 15% brine solution to give 6–8% salt in the cheese at equilibrium.

The high salt concentration retards bacterial activity. However, air should be excluded from the brining container to prevent the growth of moulds.

Feta cheese can be eaten after a few days or can be stored for long periods in the brine, provided that air is excluded. The cheese develops a soft, crumbly texture during ripening.

Expected yield: 1 kg of cheese from 9 kg of milk (11 %).

Cheese Yield

In cheese-making, the milk fat and casein are recovered with some moisture. The yield of cheese can be expressed in kilograms of cheese obtained per 100 kilograms of milk processed. Cheese yield is influenced by milk composition, the moisture content of the final cheese and the degree of recovery of fat and protein in the curd during cheese-making. Milk low in total solids will give a low cheese yield, while milk high in total solids will give a high cheese yield. In order to predict the theoretical yield of cheese, the fat and casein content of the milk must be known. Because of difficulties encountered in estimating casein content, the following formula is often used to estimate cheese yield:

$(2.3 \times \text{fat} \%) + 1.4 = \text{cheese yield (kg/ 100 kg milk)}$

Therefore, with milk containing 4% fat the expected yield would be:

$(2.3 \times 4) + 1.4 = 10.6 \text{ kg/ 100 kg milk}$

This formula gives an estimate of cheese yield and is applied most often to Cheddar cheese. It is useful as an immediate check on efficiency, but a universal yield factor for cheese varieties is unrealistic.

If the yield of cheese is less than expected, the following checks should be made:

- Weigh and record milk received.
- Sample and analyse milk received.
- Weigh, store and record cheese made.
- Sample and analyse whey.

The fat content of whey should be analysed for each batch of cheese made.

In estimating the profitability of cheese-making enterprises, an average annual yield of 9.5%, i.e. 9.5 kg of cheese per 100 kg of milk, is used.

Milk standardisation may be used to increase cheese yield, particularly with high-fat milk. Standardisation also gives a good return for skim milk. However, over-standardising results in coarse-textured cheese with poor flavour.

High moisture content increases cheese yield, but reduces keeping quality. Cheese loses moisture during storage if it is not properly wrapped, thus reducing cheese yield. Waxing reduces moisture loss, as does storing the cheese in brine.

Milk Fermentations

Raw milk produced under normal conditions develops acidity. It has long been recognised that highly acid milk does not putrefy. Therefore, allowing milk to develop acidity naturally preserves the other milk constituents.

Bacteria in milk are responsible for acid development. They produce acid by the anaerobic breakdown of milk carbohydrate—lactose—to lactic acid and other organic acids. Anaerobic breakdown of carbohydrate to organic acids or alcohols is called fermentation.

Pyruvic acid formation is an intermediate step common to most carbohydrate fermentations:

$$C_6H_{12}O_6 \longrightarrow 2\ CH_3.CO.COOH$$

However, fermentations are usually described by an identifiable end product such as lactic acid or ethyl alcohol and carbon dioxide.

A number of sugar fermentations are recognised in milk. They can be either homofermentative, with one end product, or heterofermentative, with more than one end product.

Organisms responsible:

1. Streptococci and Lactobacilli.

2. Propionibacteria.

3. Yeasts – Candida and Torula.

4. Coliform bacteria.

- The lactic acid fermentation is the most important one in milk and is central to many processes.

- Propionic fermentation is a mixed-acid fermentation and is used in the manufacture of Swiss cheese varieties.

- Alcohol fermentation can be used to prepare certain fermented milks and also to make ethyl alcohol from whey.

- The coliform gassy fermentation is an example of a spoilage fermentation. Large numbers of coliform bacteria in milk indicates poor hygiene. The coliform gassy fermentation disrupts lactic acid fermentation, and also causes spoilage in cheese.

The factors that affect microbial growth also affect milk fermentation. Fermentation rates will generally parallel the microbial growth curve up to the stationary phase. The type of fermentation obtained will depend on the numbers and types of bacteria in the milk, storage temperature and the presence or absence of inhibitory substances. The desired fermentations can be obtained by temperature manipulation or by adding a selected culture of microorganisms—starter—to pasteurised or sterilised milk.

In smallholder milk processing, traces of milk from previous batches are often used to provide 'starter' for subsequent batches. Other sources include the container and additives such as cereal grains.

The fermentation will be established once the organisms dominate the medium and will continue until either the substrate is depleted or the end product accumulates.

In milk, accumulation of end product usually arrests the fermentation. For example, accumulation of lactic acid reduces milk pH to below 4.5, which inhibits the growth of most microorganisms, including lactic-acid producers. The fermentation then slows and finally stops.

Fermented milks are wholesome foods and many have medicinal properties attributed to them.

Fermented Milks

The types of fermented milk discussed here are those made by controlled fermentation. This is achieved by establishing the desired microorganisms in the milk and by maintaining the milk at a temperature favourable to the fermentative organism.

A variety of fermented milks are made, each dithering markedly from the other. However, a number of steps are common to each manufacturing process, and these are outlined.

Standardisation

Occasionally some fat is removed or milk SNF added. In some instances, the removal of moisture during heating increases the proportion of solids in the final product.

Heating

Milk is heated to kill pathogens and spoilage organisms and to provide a cleaner medium in which the desired microorganisms can be established. Heating also removes air from the milk, resulting in a more favourable environment for the fermentative organisms, and denatures the whey proteins, which increases the viscosity of the product.

After heating, the milk must be cooled before it is inoculated with starter, otherwise the starter organisms will also be killed.

Inoculation with Starter

Starter is the term used to describe the microbial culture that is used to produce the desired fermentation and to flavour the product. When preparing the starter, care must be taken to avoid contamination with other microorganisms. Companies that supply starter cultures detail the precautions necessary. Care should also be taken to avoid contamination when inoculating the milk with starter.

Incubation

After inoculation the milk is incubated at the optimum temperature for the growth of the starter organism. Incubation is continued until the fermentation is complete, at which time the product is cooled. Additives may be added at this stage and the product packed.

Preparation of the Fermentation Vessel: The fermentation vessel is first washed to remove visible dirt. It is then dried and smoked by putting burning embers of *Olea africana,* wattle or acacia into the vessel and closing the lid. The vessel is then shaken vigorously and the lid opened to release the smoke. This procedure is repeated until the inside of the vessel is hot. Smoking flavours the product and is also thought to control the fermentation by retarding bacterial growth. While it is known that smoke contains compounds that retard bacterial growth, the precise effects of smoking on fermentation have not been investigated.

Once smoking is complete the vessel may be cleaned with a cloth to remove charcoal particles. However, in some areas the charcoal particles are retained to add colour to the product.

Milk Treatment

In some processes the milk is boiled prior to fermentation. It is then allowed to cool and the surface cream removed. In other processes the milk is not given any prefermentation treatment.

Fermentation

The milk is placed in the smoked vessel and allowed to ferment slowly in a cool place at a temperature of about 16–18°C. The fermentation is almost complete after 2 days, but may be continued for a further 2 days, by which time the flavour is fully developed. The milk must ferment at low temperature, otherwise fermentation is too vigorous, with much wheying off and gas production.

The product has a storage stability of 15 to 20 days.

Concentrated Fermented Milks

Concentrated fermented milks are prepared by removing whey from fermented milk and adding fresh milk to the residual milk constituents. The fermentation vessel is prepared as for fermented milk. The milk is allowed to ferment in a cool place for up to 7 days, during which milk may be added daily. After 7 days a coagulum has formed and the clear whey is removed. Fresh milk is then added and, following further fermentation, whey is again removed. In this way the casein and fat are gradually concentrated in a product of extended keeping quality. The actual degree of concentration depends on the amount of whey removed and of fresh milk added.

Sour-milk Technology

Smallholder milk processing is based on sour milk. This is due to a number of reasons, including high ambient temperatures, small daily quantities of milk, consumer preference and increased keeping quality of sour milk.

Products made from sour milk include fermented milks, concentrated fermented milks, butter, ghee, cottage cheese and whey. Other products made are cheese and products made by mixing fermented milk with boiled cereals. The equipment required for processing sour milk is simple and is all available locally. Milk vessels can be made from clay, gourds and wood, and can be woven from fibre, such as the *gorfu* container used by the Borana pastoralists in Ethiopia.

Butter-Making from Sour Whole Milk

This is a very important process in many parts of Africa. Smallholders produce 1 to 4 litres of milk per day for processing. Under normal storage conditions the milk becomes sour in 4 to 5 hours. The souring of milk has a number of advantages. It retards the growth of undesirable microorganisms, such as pathogens and putrefactive bacteria, and makes the milk easier to churn.

Milk for churning is accumulated over several days by adding fresh milk to the milk already accumulated. The churn holds about 20 litres and the amount of milk churned ranges from 4 to 10 litres. The milk is normally accumulated over 2 or more days. Butter is made by agitating the milk until butter grains form. The churn is then rotated slowly until the fat coalesces into a continuous mass. The butter thus formed is taken from the churn and kneaded in cold water.

The milk is usually agitated by placing the churn on a mat on the floor and rolling it to and from. It can also be agitated by shaking the churn on the lap or hung from a tripod.

A number of factors influence churning time and recovery of butterfat as butter:

- Milk acidity
- Churning temperature
- Degree of agitation, and
- Extent of filling the churn.

Effect of acidity: Fresh milk is difficult to churn: churning time is long and recovery of butterfat is poor. Milk containing at least 0.6% lactic acid is easier to churn. Acidity higher than 0.6% does not significantly influence churning time or fat recovery.

Effect of temperature: Sour milk is normally churned at between 15 and 26°C, depending on environmental temperature. At low temperatures churning time is long; butter-grain formation can take 5 hours or longer. As churning temperature increases churning time decreases. This becomes marked at temperatures above 20°C, but as little as 60% of the butterfat may be recovered as butter at 26°C. Control of temperature is therefore critical.

It is difficult to isolate the effects of temperature and acidity on churning efficiency because while the milk is ripening it is also cooling and the fat is crystallising. Direct acidification of fresh milk increases butter yield, but allowing milk to develop acidity during a ripening period of 2 to 3 days allows considerable fat crystallisation.

Degree of agitation: Increasing agitation reduces churning time. Fitting an agitator to a traditional churn reduces churning time and increases butter yield. The percentage of fat recovered as butter is increased, with as little as 0.2% fat remaining in the buttermilk. However, the process is very temperature-dependent and churning at temperatures above 20°C results in short churning times with poor recovery of fat. The optimum churning temperature is between 17 and 19°C.

Extent of filling the churn: Churns should be filled to between a third and half their volumetric capacity. Filling to more than half the volumetric capacity increases churning time considerably but does not reduce fat recovery.

Thus, when churning whole milk, the following conditions should be adhered to:
- Milk acidity should be greater than 0.6%.
- The temperature should be regulated to about 18°C.
- Internal agitation should be used to reduce churning time and increase fat recovery.
- The churn should not be filled to more than half its volumetric capacity.

Once the fat has been recovered, the soured skim milk contains casein, whey proteins, milk salts, lactic acid, lactose, the unrecovered fat and some fat-globule-membrane constituents.

Defatted milk is suitable, and is often used, for direct consumption. It is also used to inoculate fresh milk to encourage acid development.

Cottage Cheese

The casein and some of the unrecovered fat in skim milk can be heat-precipitated as cottage cheese, known in Ethiopia as *ayib*.

The defatted milk is heated to about 50°C until a distinct curd mass forms. It is then allowed to cool gradually and the curd is ladled out. Alternatively, the curd can be recovered by filtering the cooled mixture through a muslin cloth. This facilitates more complete recovery of the curd and also allows more effective moisture removal.

Temperature can be varied between 40 and 70°C without markedly affecting product composition and yield. Heat treatments between 70 and 90°C do not appear to affect yield but give the product a cooked flavour.

The whey contains about 0.75% protein, indicating near-complete recovery of casein. Whey can be consumed by humans or fed to animals.

The cottage cheese comprises 79.5% water, 14.7% protein, 1.8% fat, 0.9% ash and 3.1 % soluble milk constituents. It has a short shelf-life because of its high moisture content. Shelf-life can be increased by adding salt or by reducing the moisture content of the cheese. Storing the product in an air-tight container also extends storage life.

Equipment: Skim milk can be heated in any suitably sized vessel that is able to withstand heat. Heating can be direct or indirect. A ladle or muslin cloth can be used for product recovery.

Expected yield: The yield depends on milk composition and on the moisture content of the product, but should be at least 1 kg of cottage cheese from 8 litres of milk (12.5%).

Chapter 4

Dairy Microbiology

Basic Microbiology

Microorganisms

Microorganisms are living organisms that are individually too small to see with the naked eye. The unit of measurement used for microorganisms is the micrometer (μ m); 1 μ m = 0.001 millimetre; 1 nanometer (nm) = 0.001 μ m.

Microorganisms are found everywhere (ubiquitous) and are essential to many of our planets life processes. With regards to the food industry, they can cause spoilage, prevent spoilage through fermentation, or can be the cause of human illness.

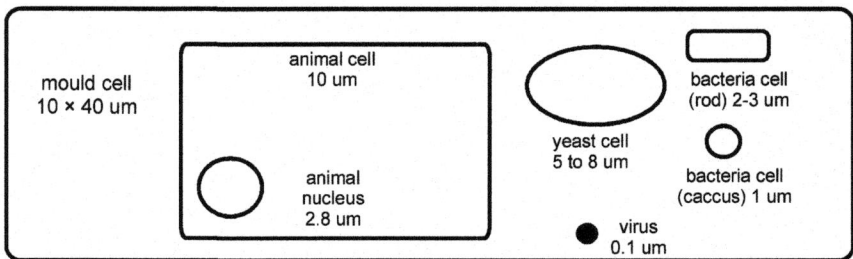

There are several classes of microorganisms, of which bacteria and fungi (yeasts and moulds) will be discussed in some detail. Another type of microorganism, the bacterial viruses or bacteriophage, will be examined in a later section.

Bacteria

Bacteria are relatively simple single-celled organisms. One method of classification is by shape or morphology:

- Cocci:

    -spherical shape

    -0.4 -1.5 μ m

Examples: staphylococci-form grape-like clusters; streptococci-form bead-like chains

• Rods:

-0.25-1.0 μ m width by 0.5-6.0 μ m long

Examples: bacilli-straight rod; spirilla-spiral rod

There exists a bacterial system of taxonomy, or classification system, that is internationally recognized with family, genera and species divisions based on genetics.

Some bacteria have the ability to form resting cells known as endospores. The spore forms in times of environmental stress, such as lack of nutrients and moisture needed for growth, and thus is a survival strategy. Spores have no metabolism and can withstand adverse conditions such as heat, disinfectants, and ultraviolet light. When the environment becomes favourable, the spore germinates and giving rise to a single vegetative bacterial cell. Some examples of spore-formers important to the food industry are members of Bacillus and Clostridium generas.

Bacteria reproduce asexually by fission or simple division of the cell and its contents. The doubling time, or generation time, can be as short as 20-20 min. Since each cell grows and divides at the same rate as the parent cell, this could under favourable conditions translate to an increase from one to 10 million cells in 11 hours! However, bacterial growth in reality is limited by lack of nutrients, accumulation of toxins and metabolic wastes, unfavourable temperatures and dessication. The maximum number of bacteria is approximately 1 X 10e9 CFU/g or ml.

Note: Bacterial populations are expressed as colony forming units (CFU) per gram or millilitre.

Bacterial growth generally proceeds through a series of phases:

• Lag phase: time for microorganisms to become accustomed to their new environment. There is little or no growth during this phase.

• Log phase: bacteria logarithmic, or exponential, growth begins; the rate of multiplication is the most rapid and constant.

• Stationary phase: the rate of multiplication slows down due to lack of nutrients and build-up of toxins. At the same time, bacteria are constantly dying so the numbers actually remain constant.

• Death phase: cell numbers decrease as growth stops and existing cells die off.

The shape of the curve varies with temperature, nutrient supply, and other growth factors. This exponential death curve is also used in modelling the heating destruction of microorganisms.

Yeasts

Yeasts are members of a higher group of microorganisms called fungi. They are single-cell organisms of spherical, elliptical or cylindrical shape. Their size varies greatly but are generally larger than bacterial cells. Yeasts may be divided into two groups according to their method of reproduction:

1. budding: called Fungi Imperfecti or false yeasts

2. budding and spore formation: called Ascomycetes or true yeasts.

Unlike bacterial spores, yeast form spores as a method of reproduction.

Moulds

Moulds are filamentous, multi-celled fungi with an average size larger than both bacteria and yeasts (10 X 40 μ m). Each filament is referred to as a hypha. The mass of hyphae that can quickly spread over a food substrate is called the mycelium. Moulds may reproduce either asexually or sexually, sometimes both within the same species.

Asexual Reproduction:

• fragmentation-hyphae separate into individual cells called arthropsores

• spore production-formed in the tip of a fruiting hyphae, called conidia, or in swollen structures called sporangium.

Sexual Reproduction: sexual spores are produced by nuclear fission in times of unfavourable conditions to ensure survival.

Microbial Growth

There are a number of factors that affect the survival and growth of microorganisms in food. The parameters that are inherent to the food, or intrinsic factors, include the following:

• nutrient content

• moisture content

• pH

• available oxygen

- biological structures
- antimicrobial constituents.

Nutrient Requirements: While the nutrient requirements are quite organism specific, the microorganisms of importance in foods require the following:

- water
- energy source
- carbon/nitrogen source
- vitamins
- minerals.

Milk and dairy products are generally very rich in nutrients which provides an ideal growth environment for many microorganisms.

Moisture Content: All microorganisms require water but the amount necessary for growth varies between species. The amount of water that is available in food is expressed in terms of water activity (aw), where the aw of pure water is 1.0. Each microorganism has a maximum, optimum, and minimum aw for growth and survival. Generally bacteria dominate in foods with high aw (minimum approximately 0.90 aw) while yeasts and moulds, which require less moisture, dominate in low aw foods (minimum 0.70 aw). The water activity of fluid milk is approximately 0.98 aw.

pH: Most microorganisms have approximately a neutral pH optimum (pH 6-7.5). Yeasts are able to grow in a more acid environment compared to bacteria. Moulds can grow over a wide pH range but prefer only slightly acid conditions. Milk has a pH of 6.6 which is ideal for the growth of many microoorganisms.

Available Oxygen: Microorganisms can be classified according to their oxygen requirements necessary for growth and survival:

- Obligate Aerobes: oxygen required
- Facultative: grow in the presence or absence of oxygen
- Microaerophilic: grow best at very low levels of oxygen
- Aerotolerant Anaerobes: oxygen not required for growth but not harmful if present
- Obligate Anaerobes: grow only in complete absence of oxygen; if present it can be lethal.

Biological Structures: Physical barriers such as skin, rinds, feathers, etc. have provided protection to plants and animals against the invasion of microorganisms. Milk, however, is a fluid product with no barriers to the spreading of microorganisms throughout the product.

Antimicrobial Constituents: As part of the natural protection against microorganisms, many foods have antimicrobial factors. Milk has several nonimmunological proteins which inhibit the growth and metabolism of many microorganisms including the following most common:

1. lactoperoxidase
2. lactoferrin
3. lysozyme
4. xanthine

Where the intrinsic factors are related to the food properties, the extrinsic factors are related to the storage environment. These would include temperature, relative humidity, and gases that surround the food.

Temperature: As a group, microorganisms are capable of growth over an extremely wide temperature range. However, in any particular environment, the types and numbers of microorganisms will depend greatly on the temperature. According to temperature, microorganisms can be placed into one of three broad groups:

- Psychrotrophs: optimum growth temperatures 20 to 30° capable of growth at temperatures less than 7° C. Psychrotrophic organisms are specifically important in the spoilage of refrigerated dairy products.
- Mesophiles: optimum growth temperatures 30 to 40° C; do not grow at refrigeration temperatures
- Thermophiles: optimum growth between 55 and 65° C

It is important to note that for each group, the growth rate increases as the temperature increases only up to an optimum, afterwhich it rapidly declines.

Detection and Enumeration of Microorganisms

There are several methods for detection and enumeration of microorganisms in food. The method that is used depends on the purpose of the testing.

Direct Enumeration: Using direct microscopic counts (DMC), Coulter counter etc. allows a rapid estimation of all viable and nonviable cells. Identification through staining and observation of morphology also possible with DMC.

Viable Enumeration: The use of standard plate counts, most probable number (MPN), membrane filtration, plate loop methos, spiral plating etc., allows the estimation of only viable cells. As with direct enumeration, these methods can be used in the food industry to enumerate fermentation, spoilage, pathogenic, and indicator organisms.

Metabolic Activity Measurement: An estimation of metabolic activity of the total cell population is possible using dye reduction tests such as resazurin or methylene blue dye reduction, acid production, electrical impedence etc. The level of bacterial activity can be used to assess the keeping quality and freshness of milk. Toxin levels can also be measured, indicating the presence of toxin producing pathogens.

Cellular Constituents Measurement: Using the luciferase test to measure ATP is one example of the rapid and sensitive tests available that will indicate the presence of even one pathogenic bacterial cell.

Isolation of microorganisms is an important preliminary step in the identification of most food spoilage and pathogenic organisms. This can be done using a simple streak plate method.

Microorganisms in Milk

Milk is sterile at secretion in the udder but is contaminated by bacteria even before it leaves the udder. Except in the case of mastisis, the bacteria at this point are harmless and few in number. Further infection of the milk by microorganisms can take place during milking, handling, storage, and other pre-processing activities.

Lactic acid bacteria: this group of bacteria are able to ferment lactose to lactic acid. They are normally present in the milk and are also used as starter cultures in the production of cultured dairy products such as yogurt. Note: many lactic acid bacteria have recently been reclassified; the older names will appear in brackets as you will still find the older names used for convenience sake in a lot of literature. Some examples in milk are:

- lactococci:
    - o *L. delbrueckii* subsp. *lactis (Streptococcus lactis)*
    - o *Lactococcus lactis* subsp. *cremoris (Streptococcus cremoris).*
- lactobacilli:
    - o *Lactobacillus casei*

o *L.delbrueckii* subsp. *lactis* (*L. lactis*)

o *L. delbrueckii* subsp. *bulgaricus* (*Lactobacillus bulgaricus*).

*   *Leuconostoc* :

Coliforms: coliforms are facultative anaerobes with an optimum growth at 37° C. Coliforms are indicator organisms; they are closely associated with the presence of pathogens but not necessarily pathogenic themselves. They also can cause rapid spoilage of milk because they are able to ferment lactose with the production of acid and gas, and are able to degrade milk proteins. They are killed by HTST treatment, therefore, their presence after treatment is indicative of contamination.*Escherichia coli* is an example belonging to this group.

Significance of microorganisms in milk::

*   Information on the microbial content of milk can be used to judge its sanitary quality and the conditions of production
*   If permitted to multiply, bacteria in milk can cause spoilage of the product
*   Milk is potentially susceptible to contamination with pathogenic microorganisms. Precautions must be taken to minimize this possibility and to destroy pathogens that may gain entrance
*   Certain microorganisms produce chemical changes that are desirable in the production of dairy products such as cheese, yogurt.

Spoilage Microorganisms in Milk

The microbial quality of raw milk is crucial for the production of quality dairy foods. Spoilage is a term used to describe the deterioration of a foods' texture, colour, odour or flavour to the point where it is unappetizing or unsuitable for human consumption. Microbial spoilage of food often involves the degradation of protein, carbohydrates, and fats by the microorganisms or their enzymes.

In milk, the microorganisms that are principally involved in spoilage are psychrotrophic organisms. Most psychrotrophs are destroyed by pasteurization temperatures, however, some like *Pseudomonas fluorescens, Pseudomonas fragi* can produce proteolytic and lipolytic extracellular enzymes which are heat stable and capable of causing spoilage.

Some species and strains of *Bacillus, Clostridium, Cornebacterium, Arthrobacter, Lactobacillus, Microbacterium, Micrococcus,* and

Streptococcus can survive pasteurization and grow at refrigeration temperatures which can cause spoilage problems.

Pathogenic Microorganisms in Milk

Hygienic milk production practices, proper handling and storage of milk, and mandatory pasteurization has decreased the threat of milkborne diseases such as tuberculosis, brucellosis, and typhoid fever. There have been a number of foodborne illnesses resulting from the ingestion of raw milk, or dairy products made with milk that was not properly pasteurized or was poorly handled causing post-processing contamination. The following bacterial pathogens are still of concern today in raw milk and other dairy products:

- *Bacillus cereus*
- *Listeria monocytogenes*
- *Yersinia enterocolitica*
- *Salmonella* spp.
- *Escherichia coli* O157:H7
- *Campylobacter jejuni.*

It should also be noted that moulds, mainly of species of *Aspergillus, Fusarium,* and *Penicillium* can grow in milk and dairy products. If the conditions permit, these moulds may produce mycotoxins which can be a health hazard.

Haccp

Raw and end-products may be tested for the presence, level, or absence of microorganisms. Traditionally these practices were used to reduce manufacturing defects in dairy products and ensure compliance with specifications and regulations, however, they have many drawbacks:

1. destructive and time consuming
2. slow response
3. small sample size
4. delays in the release of the food.

In the 1960's, the Pillsbury Company, the U.S. Army, and NASA introduced a system for assuring pathogen-free foods for the space program. This system, called Hazard Analysis and Critical Control Points (HACCP), is a focus on critical food safety areas as part of total quality programs. It involves a critical examination of the entire food manufacturing process to determine every step where there is a possibility of physical, chemical, or microbiological contamination of the food which would render it unsafe or unacceptable for human consumption. These identified points are the critical control points (CCP). There are seven prinicples to HACCP:

1. analyse hazards
2. determine CCPs
3. establish critical limits
4. establish monitoring procedures
5. establish deviation procedures
6. establish verification procedures
7. establish record keeping procedures.

Before these principles can be put into place, a prerequisite program and preliminary setup is necessary.

Prerequisite Program:

- premise control
- receiving and storage control
- equipment performance and maintenance control
- personnel training
- sanitation
- recall procedure.

Preliminary Setup:

- assemble team
- describe the product
- identify intended use
- construct flow diagram and plant schematic
- verify the diagram on-site.

Food Safety Enhancement Program-FSEP is The Canadian Food Inspection Agency's HACCP initiative. There is extensive information at their Web site regarding FSEP, including implementation manuals, HACCP curriculum guidelines, and generic models.

Starter Cultures

Starter cultures are those microorganisms that are used in the production of cultured dairy products such as yogurt and cheese. The natural microflora of the milk is either inefficient, uncontrollable, and unpredictable, or is destroyed altogether by the heat treatments given to the milk. A starter culture can provide particular characteristics in a more controlled and predictable fermentation. The primary function of lactic starters is the production of lactic acid from lactose. Other functions of starter cultures may include the following:

- flavour, aroma, and alcohol production
- proteolytic and lipolytic activities
- inhibition of undesirable organisms.

There are two groups of lactic starter cultures:

1. simple or defined: single strain, or more than one in which the number is known
2. mixed or compound: more than one strain each providing its own specific characteristics.

Starter cultures may be categorized as mesophilic or thermophilic:

Mesophilic

- *Lactococcus lactis* subsp. *cremoris*
- *L. delbrueckii* subsp. *lactis*
- *L. lactis* subsp. *lactis* biovar *diacetylactis*
- *Leuconostoc mesenteroides* subsp. *cremoris*.

Thermophilic

- *Streptococcus salivarius* subsp. *thermophilus* (*S.thermophilus*)
- *Lactobacillus delbrueckii* subsp. *bulgaricus*
- *L. delbrueckii* subsp. *lactis*
- *L. casei*
- *L. helveticus*
- *L. plantarum*.

Mixtures of mesophilic and thermophilic microorganisms can also be used as in the production of some cheeses.

Bacteriophage

Bacteriophages are viruses that require bacteria host cells for growth and reproduction. Initially, the bacteriophage attaches itself to the bacteria cell wall and injects nuclear substance into the cell. Inside the cell, the nuclear substance produces shells, or phage coats, for the new bacteriophage which are quickly filled with nucleic acid.

The bacterial cell ruptures and dies as the new bacteriophage are released.

Bacteriophages are ubiquitous but generally enter the milk processing plant with the farm milk. They can be inactivated heat treatments of 30 min at 63 to 88° C, or by the use of chemical disinfectants.

Bacteriophages are of most concern in cheese making. They attack and destroy most of the lactic acid bacteria which prevents normal ripening known as slow or dead vat.

Starter Culture Preparation

Commercial manufacturers provide starter cultures in lyophilized (freeze-dried), frozen or spray-dried forms. The dairy product manufacturers need to inoculate the culture into milk or other suitable substrate. There are a number of steps necessary for the propagation of starter culture ready for production:

1. Commercial culture
2. Mother culture-first inoculation; all cultures will originate from this preparation
3. Intermediate culture-in preparation of larger volumes of prepared starter
4. Bulk starter culture-this stage is used in dairy product production

Dairy Processing

Clarification and Cream Separation

Centrifugation: Centrifugal separation is a process used quite often in the dairy industry. Some uses include:

- clarification (removal of solid impurities from milk prior to pasteurization)
- skimming (separation of cream from skim milk)
- standardizing
- whey separation (separation of whey cream (fat) from whey)
- bactofuge treatment (separation of bacteria from milk)
- quark separation (separation of quarg curd from whey)
- butter oil purification (separation of serum phase from anhydrous milk fat).

Principles of Centrifugation

Centrifugation is based on *Stoke's Law*. The particle sedimentation velocity increases with:

- increasing diameter
- increasing difference in density between the two phases
- decreasing viscosity of the continuous phase.

If raw milk were allowed to stand, the fat globules would begin to rise to the surface in a phenomena called creaming. Raw milk in a rotating container also has centrifugal forces acting on it. This allows rapid separation of milk fat from the skim milk portion and removal of solid impurities from the milk.

Separation

Centrifuges can be used to separate the cream from the skim milk. The centrifuge consists of up to 120 discs stacked together at a 45 to 60 degree angle and separated by a 0.4 to 2.0 mm gap or separation channel. Milk is introduced at the outer edge of the disc stack. The stack of discs has vertically aligned distribution holes into which the milk is introduced.

Under the influence of centrifugal force the fat globules (cream), which are less dense than the skim milk, move inwards through the separation channels toward the axis of rotation. The skim milk will move outwards and leaves through a separate outlet.

Clarification

Separation and clarification can be done at the same time in one centrifuge. Particles, which are more dense than the continuous milk phase, are thrown back to the perimeter. The solids that collect in the centrifuge consist of dirt, epithelial cells, leucocytes, corpuscles, bacteria sediment and sludge.

The amount of solids that collect will vary, however, it must be removed from the centrifuge. More modern centrifuges are self-cleaning allowing a continuous separation/clarification process. This type of centrifuge consists of a specially constructed bowl with peripheral discharge slots.

These slots are kept closed under pressure. With a momentary release of pressure, for about 0.15 s, the contents of sediment space are evacuated. This can mean anywhere from 8 to 25 L are ejected at intervals of 60 min. For one dairy, self-cleaning translated to a loss of 50 L/hr of milk.

Standardization

The streams of skim and cream after separation must be recombined to a specified fat content. This can be done by adjusting the throttling valve of the cream outlet; if the valve is completely closed, all milk will be discharged through the skim milk outlet. As the valve is progressively opened, larger amounts of cream with diminishing fat contents are discharged from the cream outlet. With direct standardization the cream and skim are automatically remixed at the separator to provide the desired fat content.

Pasteurization

Introduction

The process of pasteurization was named after Louis Pasteur who discovered that spoilage organisms could be inactivated in wine by applying heat at temperatures below its boiling point. The process was later applied to milk and remains the most important operation in the processing of milk.

Definition: The heating of every particle of milk or milk product to a specific temperature for a specified period of time without allowing recontamination of that milk or milk product during the heat treatment process.

Purpose: There are two distinct purposes for the process of milk pasteurization:

1. Public Health Aspect-to make milk and milk products safe for human consumption by destroying all bacteria that may be harmful to health (pathogens)

2. Keeping Quality Aspect-to improve the keeping quality of milk and milk products. Pasteurization can destroy some undesirable enzymes and many spoilage bacteria. Shelf life can be 7, 10, 14 or up to 16 days.

The extent of microorganism inactivation depends on the combination of temperature and holding time. Minimum temperature and time requirements for milk pasteurization are based on thermal death time studies for the most heat resistant pathogen found in milk, *Coxelliae burnettii*. Thermal lethality determinations require the applications of microbiology to appropriate processing determinations. An overview can be found here. To ensure destruction of all pathogenic microorganisms, time and temperature combinations of the pasteurization process are highly regulated:

Ontario Pasteurization Regulations

Milk:

63° C for not less than 30 min.,

72° C for not less than 16 sec.,

or equivalent destruction of pathogens and the enzyme phosphatase as permitted by Ontario Provincial Government authorities. Milk is deemed pasteurized if it tests negative for alkaline phosphatase.

Frozen dairy dessert mix (ice cream or ice milk, egg nog):

at least 69° C for not less than 30 min;

at least 80° C for not less than 25 sec;

other time temperature combinations must be approved (e.g. 83° C/16 sec).

Milk based products-with 10% mf or higher, or added sugar (cream, chocolate milk, etc)

66° C/30 min, 75° C/16 sec

There has also been some progress with low temperature pasteurization methods using membrane processing technology..

Methods of Pasteurization

There are two basic methods, batch or continuous.

Batch Method

The batch method uses a vat pasteurizer which consists of a jacketed vat surrounded by either circulating water, steam or heating coils of water or steam.

In the vat the milk is heated and held throughout the holding period while being agitated. The milk may be cooled in the vat or removed hot after the holding time is completed for every particle. As a modification, the milk may be partially heated in tubular or plate heater before entering the vat. This method has very little use for milk but some use for milk by-products (e.g. creams, chocolate) and special batches. The vat is used extensivly in the ice cream industry for mix quality reasons other than microbial reasons.

Continuous Method

Continuous process method has several advantages over the vat method, the most important being time and energy saving. For most continuous processing, a high temperature short time (HTST) pasteurizer is used. The heat treatment is accomplished using a plate heat exchanger. This piece of equipment consists of a stack of corrugated stainless steel plates clamped together in a frame. There are several flow patterns that can be used. Gaskets are used to define the boundaries of the channels and to prevent leakage. The heating medium can be vacuum steam or hot water.

HTST Milk Flow Overview

This overview is meant as an introduction and a summary. Each piece of HTST equipment will be discussed in further detail later.

Cold raw milk at 4° C in a constant level tank is drawn into the regenerator section of pasteurizer. Here it is warmed to approximately 57° C-68° C by heat given up by hot pasteurized milk flowing in a counter current direction on the opposite side of thin, stainless steel plates. The raw milk, still under suction, passes through a positive displacement timing pump which delivers it under positive pressure through the rest of the HTST system.

The raw milk is forced through the heater section where hot water on opposite sides of the plates heat milk to a temperature of at least 72° C. The milk, at pasteurization temperature and under pressure, flows through the holding tube where it is held for at least 16 sec. The maximum velocity is governed by the speed of the timing pump, diameter and length of the holding tube, and surface friction. After passing temperature sensors of an indicating thermometer and a recorder-controller at the end of the holding tube, milk passes into the flow diversion device (FDD).

The FDD assumes a forward-flow position if the milk passes the recorder-controller at the preset cut-in temperature (>72° C). The FDD remains in normal position which is in diverted-flow if milk has not achieved preset cut-in temperature. The improperly heated milk flows through the diverted flow line of the FDD back to the raw milk constant level tank. Properly heated milk flows through the forward flow part of the FDD to the pasteurized milk regenerator section where it gives up heat to the raw product and in turn is cooled to approximately 32° C-9° C.

The warm milk passes through the cooling section where it is cooled to 4° C or below by coolant on the opposite sides of the thin, stainless steel plates. The cold, pasteurized milk passes through a vacuum breaker at least 12 inches above the highest raw milk in the HTST system then on to a storage tank filler for packaging.

Holding Time

When fluids move through a pipe, either of two distinct types of flow can be observed. The first is known as turbulent flow which occurs at high velocity and in which eddies are present moving in all directions and at all angles to the normal line of flow. The second type is streamline, or laminar flow which occurs at low velocities and shows no eddy currents. The *Reynolds number,* is used to predict whether laminar or turbulent flow will exist in a pipe:

Re < 2100 laminar

Re > 4000 fully developed turbulent flow

There is an impact of these flow patterns on holding time calculations and the assessment of proper holding tube lengths.

The holding time is determined by timing the interval for an added trace substance (salt) to pass through the holder. The time interval of the fastest particle of milk is desired. Thus the results found with water are converted to the milk flow time by formulation since a pump may not deliver the same amount of milk as it does water.

Note: the formulation assumes flow patterns are the same for milk and water. If they are not, how would this affect the efficiency of the pasteurization process?

Pressure Differential

For contiunuous pasteurizing, it is important to maintain a higher pressure on the pasteurized side of the heat exchanger. By keeping the pasteurized milk at least 1 psi higher than raw milk in regenerator, it prevents contamination of pasteurized milk with raw milk in event that a pin-hole leak develops in thin stainless steel plates. This pressure differential is maintained using a timing pump in simple systems, and differential pressure controllers and back pressure flow regulators at the chilled pasteurization outlet in more complex systems. The position of the timing pump is crucial so that there is suction on the raw regenerator side and pushes milk under pressure through pasteurized regenerator. There are several other factors involved in maintaining the pressure differential:

- The balance tank overflow level must be less than the level of lowest milk passage in the regenerator
- Properly installed booster pump is all that is permitted between balance tank and raw regenerator

- No pump after pasteurized milk outlet to vacuum breaker
- There must be greater than a 12 inch vertical rise to the vacuum breaker
- The raw regenerator drains freely to balance tank at shutdown.

Basic Component Equipment of HTST Pasteurizer

Balance Tank

The balance, or constant level tank provides a constant supply of milk. It is equipped with a float valve assembly which controls the liquid level nearly constant ensuring uniform head pressure on the product leaving the tank. The overflow level must always be below the level of lowest milk passage in regenerator. It, therefore, helps to maintain a higher pressure on the pasteurized side of the heat exchanger. The balance tank also prevents air from entering the pasteurizer by placing the top of the outlet pipe lower than the lowest point in the tank and creating downward slopes of at least 2%. The balance tank provides a means for recirculation of diverted or pasteurized milk.

Regenerator

Heating and cooling energy can be saved by using a regenerator which utilizes the heat content of the pasteurized milk to warm the incoming cold milk. Its efficiency may be calculated as follows:

- regeneration = temp. increase due to regenerator/total temp. increase

For example: Cold milk entering system at 4° C, after regeneration at 65° C, and final temperature of 72° C would have an 89.7% regeneration:

$$\frac{65-4}{72-4} = 89.7.$$

Timing Pump

The timing pump draws product through the raw regenerator and pushes milk under pressure through pasteurized regenerator. It governs the rate of flow through the holding tube. It must be a positive displacement pump equipped with variable speed drive that can be legally sealed at the maximum rate to give minimum holding time in holding tubes.

It also must be interwired so it only operates when FDD is fully forward or fully diverted, and must be "fail-safe". *A centrifugal pump with magnetic flow meter and controller may also be used.*

Holding Tube

Must slope upwards 1/4"/ft. in direction of flow to eliminate air entrapment so nothing flows faster at air pocket restrictions.

Indicating Thermometer

The indicating thermometer is considered the most accurate temperature measurement.

It is the official temperature to which the safety thermal limit recorder (STLR) is adjusted. The probe should sit as close as possible to STLR probe and be located not greater than 18 inches upstream of the flow diversion device.

Recorder-controller (STLR)

The STLR records the temperature of the milk and the time of day. It monitors, controls and records the position of the flow diversion device (FDD) and supplies power to the FDD during forward flow.

There are both pneumatic and electronic types of controllers. The operator is responsible for recording the date, shift, equipment, ID, product and amount, indicating thermometer temperature, cleansing cycles, cut in and cut out temperatures, any connects for unusual circumstances, and his/her signature.

Flow Diversion Device (FDD)

Also called the flow diversion valve (FDV), it is located at the downstream end of the upward sloping holding tube. It is essentially a 3-way valve, which, at temperatures greater than 72° C, opens to forward flow. This step requires power. At temperatures less than 72° C, the valve recloses to the normal position and diverts the milk back to the balance tank. It is important to note that the FDD operates on the measured temperature, not time, at the end of the holding period. There are two types of FDD:

- single stem-an older valve system that has the disadvantage that it can't be cleaned in place.
- dual stem-consists of 2 valves in series for additional fail safe systems. This FDD can be cleaned in place and is more suited for automation.

Vacuum Breaker

At the pasteurized product discharge is a vacuum breaker which breaks to atmospheric pressure. It must be located greater than 12 inches above the highest point of raw product in system. It ensures that nothing downstream is creating suction on the pasteurized side.

Auxiliary Equipment

Booster Pump

It is centrifugal "stuffing" pump which supplies raw milk to the raw regenerator for the balance tank. It must be used in conjunction with pressure differential controlling device and shall operate only when timing pump is operating, proper pressures are achieved in regenerator, and system is in forward flow.

Homogenizer

The homogenizer may be used as timing pump. It is a positive pressure pump; if not, then it cannot supplement flow. Free circulation from outlet to inlet is required and the speed of the homogenizer must be greater than the rate of flow of the timing pump.

Magnetic Flow Meter and Centrifugal Pump Arrangements

Magnetic flow meters can be used to measure the flow rate. It is essentially a short piece of tubing (approximately 25 cm long) surrounded by a housing, inside of which are located coils that generate a magnetic field. When milk passes through the magnetic field, it causes a voltage to be induced, and the generated signal is directly proportional to velcoity. Application of the magnetic flow meter in the dairy industry has centred around its replacing the positive displacement timing pump as the metering device in HTST pasteurizing systems, where with certain products the timing pump rotors reportedly wear out in a relatvely short period of time. In operation, the electrical signal is sent by the magnetic flow meter to the flow controller, which determines what the actual flow is compared to the flow rate set by the operator. Since the magnetic flow meter continuously senses flow rate, it will signal the electronic controller if the actual flow exceeds the set flow rate for any reason. If the flow rate is exceeded for any reason, the flow diversion device is put into diverted flow. A significant difference from the normal HTST system (with timing pump) comes into focus at this point. This system can be operated at a flow rate greater than (residence time less than) the legal limit. However, it will be in diverted flow and never in forward flow.

Another magnetic flow meter based system with an AC variable frequency motor control drive on a centrifugal pump is also possible in lieu of a positive displacement metering pump on a HTST pasteurizer. This system does not use a control valve but rather the signal from the magnetic flow meter is transmitted to the AC variable frequency control to vary the speed of the centrifugal pump. The pump, then controls the flow rate of product through the system and its holding time in the holding tube.

Automated Public Health Controllers These systems are used for time and temperature control of HTST systems. There are concerns that with sequential control, the critical control points (CCP's) are not monitored all the time; if during the sequence it got held up, the CCP's would not be monitored. With operator control, changes can be made to the program which might affect CCP's; the system is not easily sealed. No computer program can be written completely error free in large systems; as complexity increases, so too do errors. This gives rise to a need for specific regulations or computer controlled CCP's of public health significance:

1. dedicated computer-no other assignments, monitor all CCP's at least once/sec

2. not under control of any other computer system or override system, i.e., network

3. separate computer on each pasteurizer

4. I/O bus for outputs only, to other computers no inputs from other computers

5. on loss of power-public health computers should revert to fail safe position (e.g. divert)

6. last state switches during power up must be fail safe position

7. programs in ROM-tapes/disks not acceptable

8. inputs must be sealed, modem must be sealed, program sealed

9. no operator override switches

10. proper calibration procedure during that printing-Public health computer must not leave public health control for > 1 sec and upon return must complete 1 full cycle before returning to printing

11. FDV position must be monitored and temperature in holding tube recorded during change in FDV position

12. download from ROM to RAM upon startup

13. integrated with CIP computer which can be programmed e.g., FDV, booster pump controllable by CIP computer when in CIP made only.

UHT Processing

While pasteurization conditions effectively eliminate potential pathogenic microorganisms, it is not sufficient to inactivate the thermoresistant spores in milk. The term sterilization refers to the complete elimination of all microorganisms. The food industry uses the more realistic term "commercial sterilization"; a product is not necessarily free of all microorganisms, but those that survive the sterilization process are unlikely to grow during storage and cause product spoilage.

In canning we need to ensure the "cold spot" has reached the desired temperature for the desired time. With most canned products, there is a low rate of heat penetration to the thermal centre. This leads to overprocessing of some portions, and damage to nutritional and sensory characteristics, especially near the walls of the container. This implies long processing times at lower temperatures.

Milk can be made commercially sterile by subjecting it to temperatures in excess of 100° C, and packaging it in air-tight containers. The milk may be packaged either before or after sterilization. The basis of UHT, or ultra-high temperature, is the sterilization of food before packaging, then filling into pre-sterilized containers in a sterile atmosphere. Milk that is processed in this way using temperatures exceeding 135° C, permits a decrease in the necessary holding time (to 2-5 s) enabling a continuous flow operation.

Some examples of food products processed with UHT are:
*   liquid products-milk, juices, cream, yoghurt, wine, salad dressings
*   foods with discrete particles-baby foods; tomato products; fruits and vegetables juices; soups
*   larger particles-stews.

Advantages of UHT

High quality: The D and Z valves are higher for quality factors than microorganisms. The reduction in process time due to higher temperature (UHTST) and the minimal come-up and cool-down time leads to a higher quality product.

Long shelf life: Greater than 6 months, without refrigeration, can be expected.

Packaging size: Processing conditions are independent of container size, thus allowing for the filling of large containers for food-service or sale to food manufacturers (aseptic fruit purees in stainless steel totes).

Cheaper packaging: Both cost of package and storage and transportation costs; laminated packaging allows for use of extensive graphics

Difficulties with UHT

Sterility: Complexity of equipment and plant are needed to maintain sterile atmosphere between processing and packaging (packaging materials, pipework, tanks, pumps); higher skilled operators; sterility must be maintained through aseptic packaging

Particle Size: With larger particulates there is a danger of overcooking of surfaces and need to transport material-both limits particle size

Equipment: There is a lack of equipment for particulate sterilization, due especially to settling of solids and thus overprocessing

Keeping Quality: Heat stable lipases or proteases can lead to flavour deterioration, age gelation of the milk over time-nothing lasts forever! There is also a more pronounced cooked flavour to UHT milk.

UHT Methods

There are two principal methods of UHT treatment:

1. Direct Heating
2. Indirect Heating.

Direct Heating Systems

The product is heated by direct contact with steam of potable or culinary quality. The main advantage of direct heating is that the product is held at the elevated temperature for a shorter period of time. For a heat-sensitive product such as milk, this means less damage.

There are two methods of direct heating;

1. injection
2. infusion.

Injection: High pressure steam is injected into pre-heated liquid by a steam injector leading to a rapid rise in temperature. After holding, the product is flash-cooled in a vacuum to remove water equivalent to amount of condensed steam used. This method allows fast heating and cooling, and volatile removal, but is only suitable for some products. It is energy intensive and because the product comes in contact with hot equipment, there is potential for flavour damage.

Infusion: The liquid product stream is pumped through a distributing nozzle into a chamber of high pressure steam. This system is characterized by a large steam volume and a small product volume, distributed in a large surface area of product. Product temperature is accurately controlled via pressure. Additional holding time may be accomplished through the use of plate or tubular heat exchangers, followed by flash cooling in vacuum chamber. This method has several advantages:

- instantaneous heating and rapid cooling
- no localized overheating or burn-on
- suitable for low and higher viscosity products.

Indirect Heating Systems

The heating medium and product are not in direct contact, but separated by equipment contact surfaces. Several types of heat exchangers are applicable:

- plate
- tubular
- scraped surface.

Plate Heat Exchangers: Similar to that used in HTST but operating pressures are limited by gaskets. Liquid velocities are low which could lead to uneven heating and burn-on. This method is economical in floor space, easily inspected, and allows for potential regeneration.

Tubular Heat Exchangers: There are several types:

- shell and tube
- shell and coil
- double tube
- triple tube.

All of these tubular heat exchangers have fewer seals involved than with plates. This allows for higher pressures, thus higher flow rates and higher temperatures. The heating is more uniform but difficult to inspect.

Scraped Surface Heat Exchangers: The product flows through a jacketed tube, which contains the heating medium, and is scraped from the sides with a rotating knife. This method is suitable for viscous products and particulates (< 1 cm) such as fruit sauces, and can be adjusted for different products by changing configuration of rotor. There is a problem with larger particulates; the long process time for particulates would mean long holding sections which are impractical. This may lead to damaged solids and overprocessing of sauce.

Packaging for Aseptic Processing

The most important point to remember is that it must be sterile! All handling of product post-process must be within the sterile environment. There are 5 basic types of aseptic packaging lines:

1. Fill and seal: preformed containers made of thermoformed plastic, glass or metal are sterilized, filled in aseptic environment, and sealed

2. Form, fill and seal: roll of material is sterilized, formed in sterile environment, filled, sealed e.g. tetrapak

3. Erect, fill and seal: using knocked-down blanks, erected, sterilized, filled, sealed. e.g. gable-top cartons, cambri-bloc

4. Thermoform, fill, sealed roll stock sterilized, thermoformed, filled, sealed aseptically. e.g. creamers, plastic soup cans

5. Blow mold, fill, seal:

There are several different package forms that are used in aseptic UHT processing:

- cans
- paperboard/plastic/foil/plastic laminates
- flexible pouches
- thermoformed plastic containers
- flow moulded containers
- bag-in-box
- bulk totes.

It is also worth mentioning that many products that are UHT heat treated are not aseptically packaged. This gives them the advantage of a longer shelf life at refrigeration temperatures compared to pasteurization, but it does not produce a shelf-stable product at ambient temperatures, due to the possibility of recontamination post-processing.

Homogenization of Milk and Milk Products

Milk is an oil-in-water *emulsion*, with the fat globules dispersed in a continuous skimmilk phase. If raw milk were left to stand, however, the fat would rise and form a cream layer. Homogenization is a mechanical treatment of the fat globules in milk brought about by passing milk under high pressure through a tiny orifice, which results in a decrease in the average diameter and an increase in number and surface area, of the fat globules. The net result, from a practical view, is a much reduced tendency for creaming of fat globules.

Three factors contribute to this enhanced stability of homogenized milk: a decrease in the mean diameter of the fat globules (a factor in Stokes Law), a decrease in the size distribution of the fat globules (causing the speed of rise to be similar for the majority of globules such that they don't tend to cluster during creaming), and an increase in density of the globules (bringing them closer to the continuous phase) oweing to the adsorption of a protein membrane. In addition, heat pasteurization breaks down the cryo-globulin complex, which tends to cluster fat globules causing them to rise.

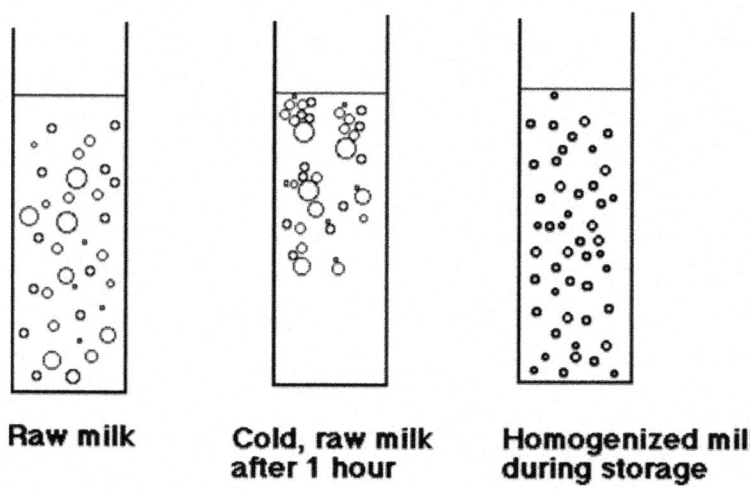

Raw milk   **Cold, raw milk**   **Homogenized mill**
                      after 1 hour   **during storage**

Homogenization Mechanism

Auguste Gaulin's patent in 1899 consisted of a 3 piston pump in which product was forced through one or more hair like tubes under pressure. It was discovered that the size of fat globules produced were

500 to 600 times smaller than tubes. There have been over 100 patents since, all designed to produce smaller average particle size with expenditure of as little energy as possible. The homogenizer consists of a 3 cylinder positive piston pump (operates similar to car engine) and homogenizing valve. The pump is turned by electric motor through connecting rods and crankshaft.

To understand the mechanism, consider a conventional homogenizing valve processing an emulsion such as milk at a flow rate of 20,000 l/hr. at 14 MPa (2100 psig). As it first enters the valve, liquid velocity is about 4 to 6 m/s. It then moves into the gap between the valve and the valve seat and its velocity is increased to 120 meter/sec in about 0.2 millisec.

The liquid then moves across the face of the valve seat (the land) and exits in about 50 microsec. The homogenization phenomena is completed before the fluid leaves the area between the valve and the seat, and therefore emulsification is initiated and completed in less than 50 microsec.

The whole process occurs between 2 pieces of steel in a steel valve assembly. The product may then pass through a second stage valve similar to the first stage. While most of the fat globule reduction takes place in the first stage, there is a tendency for clumping or clustering of the reduced fat globules. The second stage valve permits the separation of those clusters into individual fat globules.

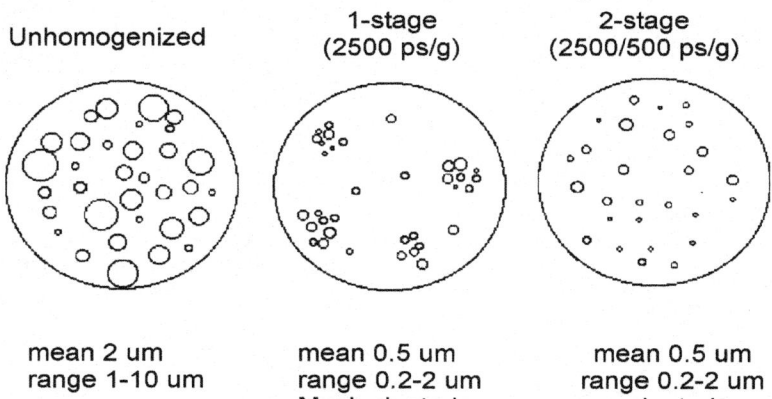

Unhomogenized	1-stage (2500 ps/g)	2-stage (2500/500 ps/g)
mean 2 um range 1-10 um	mean 0.5 um range 0.2-2 um Much clustering	mean 0.5 um range 0.2-2 um noclustering

Figure : The Effects of 2-stage Homogenization on Fat Globule Size Distribution as seen under the light microscope.

It is most likely that a combination of two theories, turbulence and cavitation, explains the reduction in size of the fat globules during the homogenization process.

Turbulence

Energy, dissipating in the liquid going through the homogenizer valve, generates intense turbulent eddies of the same size as the average globule diameter. Globules are thus torn apart by these eddie currents reducing their average size.

Cavitation

Considerable pressure drop with charge of velocity of fluid. Liquid cavitates because its vapour pressure is attained. Cavitation generates further eddies that would produce disruption of the fat globules. The high velocity gives liquid a high kinetic energy which is disrupted in a very short period of time. Increased pressure increases velocity. Dissipation of this energy leads to a high energy density (energy per volume and time). Resulting diameter is a function of energy density.

In summary, the homogenization variables are:

- type of valve
- pressure
- single or two-stage
- fat content
- surfactant type and content
- viscosity
- temperature.

Also to be considered are the droplet diameter (the smaller, the more difficult to disrupt), and the log diameter which decreases linearly with log P and levels off at high pressures.

Effect of Homogenization

Fat Globule

	No Homogenization	15 MPa (2500 psig)
Av. diam. (μ m)	3.3	0.4
Max. diam. (μ m)	10	2
Surf. area (m²/ml of milk)	0.08	0.75
Number of globules (μ m^{-3})	0.02	12

Surface Layer

The milk fat globule has a native membrane, picked up at the time of secretion, made of amphiphilic molecules with both hydrophilic and hydrophobic sections. This membrane lowers the interfacial tension resulting in a more stable emulsion. During homogenization, there is a tremendous increase in surface area and the native milk fat globule membrane (MFGM) is lost. However, there are many amphiphilic molecules present from the milk plasma that readily adsorb: casein micelles (partly spread) and whey proteins.

The interfacial tension of raw milk is 1-2 mN/m, immediately after homogenization it is unstable at 15 mN/m, and shortly becomes stable (3-4 mN/m) as a result of the adsorption of protein. The transport of proteins is not by diffusion but mainly by convection. Rapid coverage is achieved in less than 10 sec but is subject to some rearrangement.

Surface excess is a measure of how much protein is adsorbed; for example 10 mg/m² translates to a thickness of adsorbed layer of approximately 15 nm.

Membrane Processing

Membrane processing is a technique that permits concentration and separation without the use of heat. Particles are separated on the basis of their molecular size and shape with the use of pressure and specially designed semi-permeable membranes. There are some fairly new developments in terms of commercial reality and is gaining readily in its applications:

- proteins can be separated in whey for the production of whey protein concentrate (WPC)
- milk can be concentrated prior to cheesemaking at the farm level
- apple juice and wine can be clarified
- waste treatment and product recovery is possible in edible oil, fat, potato, and fish processing
- fermentation broths can be clarified and separated
- whole egg and egg white ultrafiltration as a preconcentration prior to spray drying.

Principle of Operation

When a solution and water are separated by a semi-permeable membrane, the water will move into the solution to equilibrate the system. This is known as *osmotic pressure* If a mechanical force is applied to exceed the osmotic pressure (up to 700 psi), the water is forced to move down the concentration gradient i.e. from low to high concentration. Permeate designates the liquid passing through the membrane, and retentate (concentrate) designates the fraction not passing through the membrane.

Membrane Processing

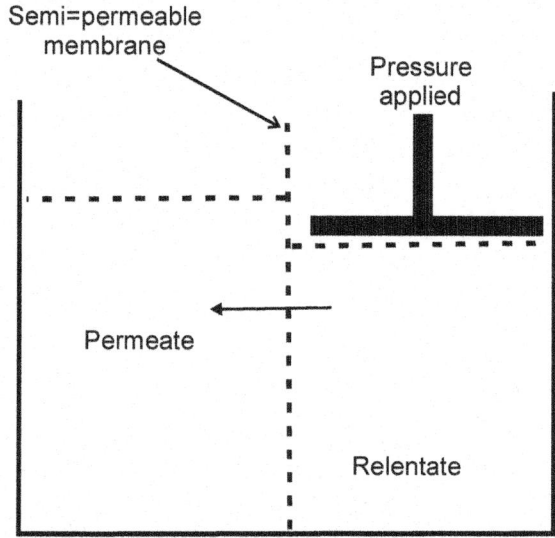

Membrane Processing

Reverse Osmosis

Reverse osmosis (RO) designates a membrane separation process, driven by a pressure gradient, in which the membrane separates the solvent (generally water) from other components of a solution. The membrane configuration is usually cross-flow. With reverse osmosis, the membrane pore size is very small allowing only small amounts of very low molecular weight solutes to pass through the membranes. It is a concentration process using a 100 MW cutoff, 700 psig, temperatures less than 40°C with cellulose acetate membranes and 70-80°C with composite membranes.

Hyperfiltration is the same as RO.

Ultrafiltration

Ultrafiltration (UF) designates a membrane separation process, driven by a pressure gradient, in which the membrane fractionates components of a liquid as a function of their solvated size and structure. The membrane configuration is usually cross-flow. In UF, the membrane pore size is larger allowing some components to pass through the pores with the water. It is a separation/ fractionation process using a 10,000 MW cutoff, 40 psig, and temperatures of 50-

60°C with polysulfone membranes. In UF milk, lactose and minerals pass in a 50% separation ratio; for example, in the retentate would be 100% of fat, 100% of protein, 50% of lactose, and 50% of free minerals.

Diafiltration is a specialized type of ultrafiltration process in which the retentate is diluted with water and re-ultrafiltered, to reduce the concentration of soluble permeate components and increase further the concentration of retained components.

Microfiltration

Microfiltration (MF) designates a membrane separation process similar to UF but with even larger membrane pore size allowing particles in the range of 0.2 to 2 micrometers to pass through. The pressure used is generally lower than that of UF process. The membrane configuration is usually cross-flow. MF is used in the dairy industry for making low-heat sterile milk as proteins may pass through but bacteria do not.

Hardware Design

Open Tubular

Tubes of membrane with a diameter of 1/2 to 1 inch and length to 12 ft. are encased in reinforced fibreglass or enclosed inside a rigid PVC or stainless steel shell. As the feed solution flows through the membrane core, the permeate passes through the membrane and is collected in the tubular housing. Imagine 12 ft long straws!

Hollow Fibre

Similar to open tubular, but the cartridges contain several hundred very small (1 mm diam) hollow membrane tubes or fibres. As the feed solution flows through the open cores of the fibres, the permeate is collected in the cartridge area surrounding the fibres.

Plate and Frame

This system is set up like a plate heat exchanger with the retentate on one side and the permeate on the other. The permeate is collected through a central collection tube.

Spiral Wound

This design tries to maximize surface area in a minimum amount of space. It consists of consecutive layers of large membrane and support material in an envelope type design rolled up around a perforated steel tube.

Electrodialysis

Electrodialysis is used for demineralization of milk products and whey for infant formula and special dietary products. Also used for desalination of water.

Principles of Operation

Under the influence of an electric field, ions move in an aqueous solution. The ionic mobility is directly proportioned to specific conductivity and inversely proportioned to number of molecules in solution. ~3-6 x 102 mm/sec.

Charged ions can be removed from a solution by synthetic polymer membranes containing ion exchange groups. Anion exchange membranes carry cationic groups which repel cations and are permeable to anions, and cation exchange membranes contain anionic groups and are permeable only to cations.

Electrodialysis membranes are comprised of polymer chains-styrene-divinyl benzene made anionic with quaternary ammonium groups and made cationic with sulphonic groups. 1-2V is then applied across each pair of membranes.

Electrodialysis Process

Amion and cation exchange membranes are arranged alternately in parallel between an anode and a cathode. The distance between the membranes is 1mm or less. A plate and frame arrangement similar to a plate heat exchanger or a plate filter is used. The solution to be demineralized flows through gaps between the two types of membranes. Each type of membrane is permeable to only one type of ion. Thus, the anions leave the gap in the direction of the anode and cations leave in the direction of the cathode. Both are then taken up by a concentrating stream.

Problems

Concentration polarization. Deposits on membrane surfaces, e.g. proteins-pH control is important. Prior concentration of whey, to 20% TS, is necessary before electrodialysis.

Ion Exchange

Ion exchange is not a membrane process but I have included it here anyway because it is used for product of protein isolates of higher concentration than obtainable by membrane concentration. Fractionation may also be accomplished using ion exchange processing.

It relies on inert resins (cellulose or silica based) that can adsorb charged particles at either end of the pH scale. The design can be a batch type, stirred tank or continuous column. The column is more suitable for selective fractionation. Whey protein isolate (WPI), with a 95% protein content, can be produced by this method. Following adsorption and draining of the deproteined whey, the pH or charge properties are altered and proteins are eluted. Protein is recovered from the dilute stream through UF and drying. Selective resins may be used for fractionated protein products or enriched in fraction allow tailoring of ingredients.

Evaporation and Dehydration

The removal of water from foods provides microbiological stability, reduces deteriorative chemical reactions, and reduces transportation and storage costs. Both evaporation and dehydration are methods used in the dairy industry for this purpose.

Evaporation

Evaporation refers to the process of heating liquid to the boiling point to remove water as vapour. Because milk is heat sensitive, heat damage can be minimized by evaporation under vacuum to reduce the boiling point. The basic components of this process consist of:

*   heat-exchanger
*   vacuum
*   vapour separator
*   condenser.

The heat exchanger is enclosed in a large chamber and transfers heat from the heating medium, usually low pressure steam, to the product usually via indirect contact surfaces. The vacuum keeps the product temperature low and the difference in temperatures high. The vapour separator removes entrained solids from the vapours, channelling solids back to the heat exchanger and the vapours out to the condenser. It is sometimes a part of the actual heat exchanger, especially in older vacuum pans, but more likely a separate unit in newer installations. The condenser condenses the vapours from inside the heat exchanger and may act as the vacuum source.

Principle of Operation

The driving force for heat transfer is the difference in temperature between the steam in the coils and the product in the pan. The steam is produced in large boilers, generally tube and chest heat exchangers.

The steam temperature is a function of the steam pressure. Water boils at 100° C at 1 atm., but at other pressures the boiling point changes. At its boiling point, the steam condenses in the coils and gives up its latent heat. If the steam temperature is too high, burn-on/fouling increases so there are limits to how high steam temperatures can go. The product is also at its boiling point. The boiling point can be elevated with an increase in solute concentration. This *boiling point elevation* works on the same principles as freezing point depression.

Evaporator Designs

Types of single effect evaporators:
- Batch Pan
- Rising film
- Falling film
- Plate evaporators
- Scraped surface.

Batch pan evaporators are the simplest and oldest. They consist of spherical shaped, steam jacketed vessels. The heat transfer per unit volume is small requiring long residence times. The heating is due only to natural convection, therefore, the heat transfer characteristics are poor. Batch plants are of historical significance; modern evaporation plants are far-removed from this basic idea. The vapours are a tremendous source of low pressure steam and must be reused.

Rising film evaporators consist of a heat exchanger isolated from the vapour separator. The heat exchanger, or calandria, consists of 10 to 15 meter long tubes in a tube chest which is heated with steam. The liquid rises by percolation from the vapours formed near the bottom of the heating tubes. The thin liquid film moves rapidly upwards. The product may be recycled if necessary to arrive at the desired final concentration. This development of this type of modern evaporator has given way to the falling film evaporator.

The falling film evaporators are the most widely used in the food industry. They are similar in components to the rising film type except that the thin liquid film moves downward under gravity in the tubes. A uniform film distribution at the feed inlet is much more difficult to obtain. This is the reason why this development came slowly and it is only within the last decade that falling film has superceded all other designs. Specially designed nozzles or spray distributors at the feed inlet permit it to handle more viscous products. The residence time is 20-30 sec. as opposed to 3-4 min. in the rising film type. The vapour separator is at the bottom which decreases the product hold-up during shut down. The tubes are 8-12 meters long and 30-50 mm in diameter.

Multiple Effect Evaporators

Two or more evaporator units can be run in sequence to produce a multiple effect evaporator. Each effect would consist a heat transfer surface, a vapour separator, as well as a vacuum source and a condenser. The vapours from the preceding effect are used as the heat source in the next effect. There are two advantages to multiple effect evaporators:

- economy-they evaporate more water per kg steam by re-using vapours as heat sources in subsequent effects
- improve heat transfer-due to the viscous effects of the products as they become more concentrated

Each effect operates at a lower pressure and temperature than the effect preceding it so as to maintain a temperature difference and continue the evaporation procedure. The vapours are removed from the preceding effect at the boiling temperature of the product at that effect so that no temperature difference would exist if the vacuum were not increased. The operating costs of evaporation are relative to the number of effects and the temperature at which they operate. The boiling milk creates vapours which can be recompressed for high steam economy. This can be done by adding energy to the vapour in the form of a steam jet, thermo compression or by a mechanical compressor, mechanical vapour recompression.

Thermo Compression (TC)

Involves the use of a steam-jet booster to recompress part of the exit vapours from the first effect. Through recompression, the pressure and temperature of the vapours are increased. As the vapours exit from the first effect, they are mixed with very high pressure steam. The steam entering the first effect calandria is at slightly less pressure than the supply steam. There is usually more vapours from the first effect than the second effect can use; usually only the first effect is coupled with multiple effect evaporators.

Mechanical Vapour Recompression (MVR)

Whereas only part of the vapour is recompressed using TC, all the vapour is recompressed in an MVR evaporator. Vapours are mechanically compressed by radial compressors or simple fans using electrical energy.

There are several variations; in single effect, all the vapours are recompressed therefore no condensing water is needed; in multiple effect, can have MVR on first effect, followed by two or more traditional effects; or can recompress vapours from all effects.

Dehydration

Dehydration refers to the nearly complete removal of water from foods to a level of less than 5%. Although there are many types of driers, spray driers are the most widely used type of air convection drier. It turns out more tonnage of dehydrated products than all other types of driers combined. It is limited to food that can be atomized, i.e. liquids, low viscosity pastes, and purees. Drying takes place within a matter of seconds at temperatures approximately 200° C. Evaporative cooling maintains low product temperatures, however, prompt removal of the product is still necessary.

Spray Drying-Process—Summary

The liquid food is generally preconcentrated by evaporation to economically reduce the water content. The concentrate is then introduced as a fine spray or mist into a tower or chamber with heated air. As the small droplets make intimate contact with the heated air, they flash off their moisture, become small particles, and drop to the bottom of the tower and are removed. The advantages of spray drying include a low heat and short time combination which leads to a better quality product.

Principal components include:

- a high pressure pump for introducing liquid into the tower
- a device for atomizing the feed stream
- a heated air source with blower
- a secondary collection vessel for removing the dried food from the airstream
- means for exhausting the moist air
- usually includes a preconcentration step i.e. MVR evaporation.

Atomizing devices are the distinguishing characteristic of spray drying. They provide a large surface area for exposure to drying forces:

1 litre = 12 billion particles = >300 ft^2 (30m^2)

The exit air temperature is an important parameter to monitor because it responds readily to changes in the process and reflects the quality of the product. Generally, we want it high enough to yield desired moisture without heat damage. There are two controls that may be used to adjust the exit air temperature:

- altering feed flow rate
- altering inlet temperature.

If heat damage occurs before the product is dried, the particle size must be reduced; smaller particle dries faster, therefore, less heat damage. This can be accomplished in three ways:

- smaller orifice
- increase atomizing pressure
- reduce viscosity-by increasing feed temperature or reducing solids.

Powder Recovery

It is essential for both economic and environmental reasons that as much powder as possible be recovered from the air stream. Three systems are available, however wet scrubbers usually act as a secondary collection system following a cyclone.

Bag Filters

Bag filters are very efficient (99.9%), but not as popular due to labour costs, sanitation, and possible heat damage because of the long residence time. They are not recommended in the case of handling high moisture loads or hygroscopic particles.

Cyclone Collector

Cyclones are not as efficient (99.5%) as bag filters but several can be placed in series. Air enters at tangent at high velocity into a cylinder or cone which has a much larger cross section. Air velocity is decreased in the cone permitting settling of solids by gravity. Centrifugal force is important in removing particles from the air stream. High air velocity is needed to separate small diameter and light materials from air; velocities may approach 100 ft/sec (70 MPH). Higher centrifugal force can be obtained by using small diameter cyclones, several of which may be placed in parallel; losses may range from 0.5-2%. A rotary airlock is used to remove powder from the cyclone. (An example of a rotary airlock is a revolving door at a hotel lobby which is intended to break the outside and inside environments).

Wet Scrubber

Wet scrubbers are the most economical outlet air cleaner. The principle of a wet scrubber is to dissolve any dust powder left in the airstream into either water or the feed stream by spraying the wash stream through the air. This also recovers heat from the exiting air and evaporates some of the water in the feed stream (if used as the wash water).

Wet scrubbers not only recover most of what would be lost product, but also recover approximately 90% of the potential drying energy normally lost in exit air. The exit air picks up moisture which increases evaporative capacity by 8% (concentration of feed). Cyclone separators are probably the best primary powder separator system because they are hygienic, easy to operate, and versatile, however, high losses may occur. Wet scrubbers are designed for a secondary air cleaning system in conjunction with the cyclone. Either feed stream or water can be used as scrubbing liquor. Also, there are heat recovery systems available.

Two and Three Stage Spray Drying With a Fluidized Bed

Principles of Fluid Beds

Air is blown up through a wire mesh belt on porous plate that supports and conveys the product. A slight vibration motion is imparted to the food particles. When the air velocity is increased to the point where it just exceeds the velocity of free fall (gravity) of the particles, fluidization occurs. The dancing/boiling motion subdivides the product and provides intimate contact of each particle with the air, but keeps clusters from forming.

With products that are particularly difficult to fluidize, a vibrating motion of the drier itself is used to aid fluidization; it is called vibro-fluidizer which is on springs. The fluidized solid particles then behave in an analogous manner to a liquid., i.e. they can be conveyed. Air velocities will vary with particle size and density, but are in the range of 0.3-0.75 m/s. They can be used not only for drying but also for cooling. If the velocity is too high, the particles will be carried away in the gas stream, therefore, gravitational forces need to be only slightly exceeded.

Two and Three Stage Drying Processes

In standard, single stage spray drying, the rate of evaporation is particularly high in the first part of the process, and it gradually decreases because of the falling moisture content of the particle surfaces. In order to complete the drying in one stage, a relatively high outlet temperature is required during the final drying phase. Of course the outlet temperature is reflective of the particle temperature and thus heat damage.

Consequently the two stage drying process was introduced which proved to be superior to the traditional single stage drying in terms of product quality and cost of production.

The two stage drier consists of a spray drier with an external vibrating fluid bed placed below the drying chamber. The product can be removed from the drying chamber with a higher moisture content, and the final drying takes place in the external fluid bed where the residence time of the product is longer and the temperature of the drying air lower than in the spray dryer.

This principle forms the basis of the development of the three stage drier. The second stage is a fluid bed built into the cone of the spray drying chamber. Thus it is possible to achieve an even higher moisture content in the first drying stage and a lower outlet air temperature from the spray drier. This fluid bed is called the integrated fluid bed. The inlet air temperature can be raised resulting in a larger temperature difference and improved efficiency in the drying process. The exhaust heat from the chamber is used to preheat the feed stream. The third stage is again the external fluid bed, which can be static or vibrating, for final drying and/or cooling the powder. The results are as follows:

- higher quality powders with much better rehydrating properties directly from the drier
- lower energy consumption
- increased range of products which can be spray dried i.e., non density, non hygroscopic
- smaller space requirements.

Agglomerating and Instantizing

These processes have allowed the manufacturing of milk powders with better reconstitution properties, such as instantized skim milk powder.

Agglomeration Mechanism: Powder is wetted with water or steam. The surface must be uniformly wetted but not excessively. The powder is held wet over a selected period of time to give moisture stability to the clusters which have formed. The clusters are dried to the desired moisture content and then cooled (e.g., fluid bed). Dried clusters are screened and sized to reduce excessively large particles and remove excessively small ones. The agglomeration process causes an increase in the amount of air incorporated between powder particles. More incorporated air is replaced with more water when the powder is reconstituted, which immediately wet the powder particles.

Agglomerating Techniques

Rewet Methods

This method uses powder as feed stock. An example is the ARCS Instantizer. Humidified air moistens powder, which causes it to cluster. It is re-dried and wetted. The clustered powder is then exposed to heated, filtered, high-velocity air. The dried clusters are then exposed to cooled air on a vibrating belt. It is then sized (pelleted to uniform size) and the fines are removed.

Straight Thru Process

A multi-stage drying process produces powders with much better solubility characteristics similar to instantized powder. This method uses a low outlet temperature which allows higher moisture in powder as it is taken from spray drier with excess moisture removed in the fluid bed. The powder fines are reintroduced to the atomizing cloud in the drying chamber.

Production and Utilization of Steam and Refrigeration

This section will describe the production and utilization of both steam and refrigeration, two utilities absolutely essential to the operation of a dairy processing plant.

Steam Production and Utilization

Understanding Steam

The diagram below helps to explain the various principles involved in the thermodynamics of steam. It shows the relationship between temperature and enthalpy (energy or heat content) of water as it passes through its phase change.

The reference point for enthalpy of water and steam is 0°C, at which point an enthalpy value of 0 kJ/kg is given to it (but of course water at 0° has alot of energy in it, which is given up as it freezes- it's not until 0K, absolute zero, when it truly has no enthalpy!). As we increase the temperature of water, its enthalpy increases by 4.18 kJ/kg °C until we hit its boiling point (which is a function of its pressure-the boiling point of water is 100°C ONLY at 1 atm. pressure). At this point, a large input of enthalpy causes no temperature change but a phase change, latent heat is added and steam is produced. Once all the water has vaporized, the temperature again increases with the addition of heat (sensible heat of the vapour).

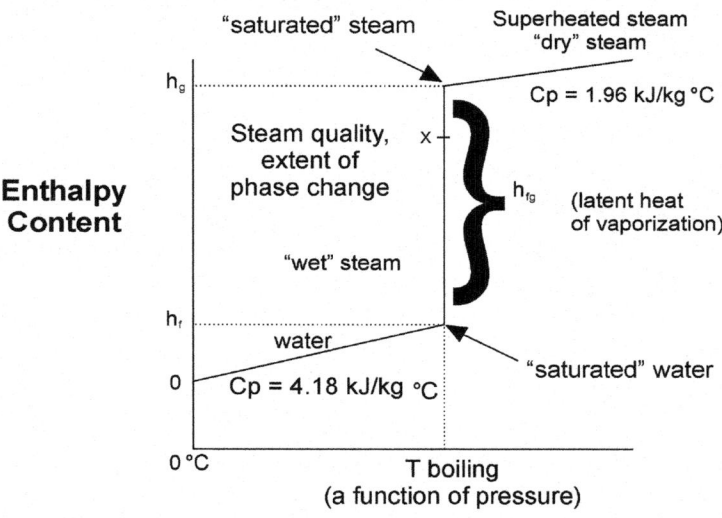

Temperature

Steam Production and Distribution

Steam is produced in large tube and chest heat exchangers, called water tube boilers if the water is in the tubes, surrounded by the flame, or fire tube boilers if the opposite is true. The pressure inside a boiler is usually high, 300-800 kPa. The steam temperature is a function of this pressure.

The steam, usually saturated or of very high quality, is then distributed to the heat exchanger where it is to be used, and it provides heat by condensing back to water (called condensate) and giving up its latent heat.

The temperature desired at the heat exchanger can be adjusted by a pressure reducing valve, which lowers the pressure to that corresponding to the desired temperature. After the steam condenses in the heat exchanger, it passes through a steam trap (which only allows water to pass through and hence holds the steam in the heat exchanger) and then the condensate (hot water) is returned to the boiler so it can be reused.

Condensed Milk and Milk Powder

Condensed Milk

Condensed milk, also known as sweetened condensed milk, is cow's milk from which water has been removed and to which sugar has been added, yielding a very thick, sweet product that can last for years without refrigeration if unopened. The two terms, condensed milk and sweetened condensed milk, have become synonymous; though there have been unsweetened condensed milk products, today these are uncommon.

History

According to the writings of Marco Polo, the Tartars were able to condense milk. Ten pounds (4.54 kg) of milk paste was carried by each man who would mix the product with water. However, this probably refers to the soft Tartar curd which can be made into a drink ("airan") by diluting it and therefore to fermented, not fresh, milk concentrate.

Nicolas Appert condensed milk in France in 1820. and was developed after in the United States in 1856 by Gail Borden, Jr. in reaction to difficulties in storing milk for more than a few hours. Before this development milk could only be kept fresh for a short while and was only available in the immediate vicinity of a cow. While returning from a trip to England in 1851, Borden was devastated by the death of several children, apparently due to poor milk from shipboard cows. With less than a year of schooling and following in a wake of failures, both of his own and of others, Borden was inspired by the vacuum pan he had seen being used by Shakers to condense fruit juice and was at last able to reduce milk without scorching or curdling it. Even then his first two factories failed and only the third, built with his new partner, Jeremiah Milbank in Wassaic, New York, produced a usable milk derivative that was long-lasting and needed no refrigeration.

Probably of equal importance for the future of milk were Borden's requirements (the "Dairyman's Ten Commandments") for farmers who wanted to sell him raw milk: they were required to wash udders before milking, keep barns swept clean, and scald and dry their strainers morning and night. By 1858 Borden's milk, sold as Eagle Brand, had gained a reputation for purity, durability and economy.

In 1864, Gail Borden's New York Condensed Milk Company constructed the New York Milk Condensery in Brewster, New York.

This condensery was the largest and most advanced milk factory and was Borden's first commercially successful plant. Over 200 dairy farmers supplied 20,000 gallons (76,000 litres) of milk daily to the Brewster plant as demand was driven by the Civil War.

The U.S. government ordered huge amounts of it as a field ration for Union soldiers during the American Civil War. This was an extraordinary field ration for the nineteenth century: a typical 14 oz (400 g) can contains 1,300 calories (5440 kj), 1 oz (30 g) each of protein and fat, and more than 7 oz (200 g) of carbohydrate.

Soldiers returning home from the Civil War soon spread the word. By the late 1860s, condensed milk was a major product. The first Canadian condensery was built at Truro, Nova Scotia, in 1871. In 1899, E. B. Stuart opened the first Pacific Coast Condensed Milk Company (later known as the Carnation Milk Products Company) plant in Kent, Washington. Unfortunately, the condensed milk market developed a bubble. Too many manufacturers chased too little demand. By 1912, stocks of condensed milk were large and the price dropped. Many condenseries went out of business. In 1911, Nestlé constructed the world's largest condensed milk plant in Dennington, Victoria, Australia.

In 1914, Professor Otto F Hunziker, head of Purdue University's dairy department, self-published *Condensed milk and milk powder: prepared for the use of milk condenseries, dairy students and pure food departments*. This text, along with additional work of Professor Hunziker and others involved with the American Dairy Science Association, standardized and improved condensery operations in the U.S. and internationally. Hunziker's book was republished in a seventh edition in October 2007 by Cartwright Press.

The first World War regenerated interest in, and a market for, condensed milk, primarily due to its storage and transportation benefits. In the U.S., the higher price for raw milk paid by condenseries created significant problems for the cheese industry.

Production

Raw milk is clarified and standardized, and then is heated to 85-90°C for several seconds. This heating destroys some microorganisms, decreases fat separation and inhibits oxidation. Some water is evaporated from the milk and sugar is added to approximately 45%. This sugar is what extends the shelf life of sweetened condensed milk. Sucrose increases the liquid's osmotic pressure, which prevents microorganism growth. The sweetened evaporated milk is cooled and lactose crystallization is induced.

Current Use

Condensed milk is used in recipes for the popular Brazilian candy brigadeiro in which condensed milk is the main ingredient (the most famous condensed milk brand in Brazil is Moça , local version of Swiss *Milch Mädchen* marketed by Nestlé), lemon meringue pie, key lime pie, caramel candies and other desserts.

In parts of Asia and Europe, sweetened condensed milk is the preferred milk to be added to coffee or sweetened tea. Many countries in South East Asia use condensed milk to flavour their coffee. A popular treat in Asia is to put condensed milk on toast and eat it in a similar way as jam and toast. Nestlé has even produced a squeeze bottle similar to Smucker's jam squeeze bottles for this very purpose. Condensed milk is a major ingredient in many Indian desserts and sweets. While most Indians start with normal milk to reduce and sweeten it, packaged condensed milk has also become popular.

In New Orleans, it is commonly used as a topping on top of a chocolate or similar cream flavour snowball. In Scotland, it is mixed with sugar and some butter and baked to form a popular, sweet candy called Tablet (confectionery) or Swiss-Milk-Tablet.

In some parts of the Southern U.S., condensed milk is a key ingredient in lemon icebox pie, a sort of cream pie. In the Philippines, condensed milk is mixed with some evaporated milk and eggs, spooned into shallow metal containers over liquid caramelised sugar, then steamed to make a stiffer and more filling version of crème brulée known as leche flan.

During the communism era in Poland it was common to boil a can of condensed milk in water for about 2 hours. The resulting product is called kaymak-sweet semiliquid substance which can be used as a cake icing or put between dry wafers. It is less common nowadays but recently some manufactures of condensed milk introduced canned ready-made kaymak. Boiling the can in this way is central to the making of Banoffee pie and home-made dulce de leche.

Substitutions

To gain condensed milk from 1 cup (250 ml) of evaporated milk one has to add 1 1/4 cups (250 g) of sugar and dissolve it by heating the milk.

Condensed Milk Companies

- Arla Foods (Canned Cream & Milk Company)
- Borden Food Corporation (New York Condensed Milk Company)
- Carnation Evaporated Milk Company (Pacific Coast Condensed Milk Company, Mohawk Milk Company)
- Dairymen's League Cooperative Association
- Foremost Condensery (Western Condensing)
- Milk Producers Cooperative Marketing Company
- Nestlé (Anglo/Swiss Condensed Milk Company)
- Northern Condensed Milk Company
- Northern Foods plc
- Pet Milk Company (Helvetia Milk Condensing Company, Utah Condensed Milk Company)
- Wisconsin Condensed Milk Company (R. G. Fraser & Company).

Baked Milk

Baked milk is a variety of boiled milk that has been particularly popular in Russia and Ukraine. It is made by cooking boiled milk on low heat for eight hours or more.

In rural areas, baked milk has been produced by leaving a jug of boiled milk in an oven for a day or for a night until it is coated with a brown crust. The stove in a traditional Russian loghouse (izba) was designed so as to "sustain varying cooking temperatures based on the placement of the food inside the oven".

Nowadays baked milk is produced on an industrial scale, as is soured or fermented baked milk, traditionally known as ryazhenka. Like scalded milk it is free of bacteria and enzymes, and so can be stored safely at room temperature for up to forty hours. Home-made baked milk was (and still is) used for preparing a range of cakes, pies, and cookies.

Caramelized condensed milk is a similar preparation used in homemade pastries. It is sometimes prepared by prolonged heating of the unopened cans with sweetened condensed milk.

Evaporated Milk

Evaporated milk, also known as dehydrated milk, is a shelf-stable canned milk product with about 60% of the water removed from fresh milk. It differs from sweetened condensed milk, which contains added sugar. Sweetened condensed milk requires less processing since the added sugar inhibits bacterial growth. The actual liquid portion of the product takes up half the space of fresh milk. When the non-liquid product is mixed with a proportionate amount of water, evaporated milk becomes the equivalent of fresh milk.

This makes evaporated milk attractive for shipping purposes and can have a shelf life of months or even years, depending upon the brand. This made evaporated milk very popular before refrigeration as a safe and reliable substitute for perishable fresh milk, that could be shipped easily to locations lacking the means of safe milk production or storage. Households in the western world use it most often today for desserts and baking due to its unique flavour. It is also used as a substitute for pouring cream, as an accompaniment to desserts, or (undiluted) as a rich substitute for milk.

History

Condensed milk was introduced to the U.S. by Gail Borden which he made using a process under the patent issued on August 19, 1856. It became popular for those people who were remote from farm sources, since it was capable of long term storage.

The invention of evaporated milk followed three decades later when John B. Meyenberg emigrated to the U.S. from Switzerland where he had devised the process, but had no support to begin production.

He obtained two U.S. patents for his process and sterilizing apparatus, issued on November 25, 1884. He formed the Helvetia Milk Condensing Company on February 14, 1885, with a number of farmers and businessmen of Highland, Illinois, as stockholders. By June 14, 1885, the first canned "Highland Evaporated Cream" was ready to be marketed. There were problems with the new product, with premature spoilage in early batches. Over the next few years, Louis Latzer and Dr. Werner Schmidt solved the problems which had been found to be caused by bacteria. With the marketing efforts of John Wilde, the company became successful as Pet, Inc., and is now part of The J.M. Smucker Co.

John P. Meyenberg, son of John B. Meyenberg, was the first American to evaporate goat's milk. He started the Meyenberg business in 1934, supplying goat milk products that are more digestible than cow's milk, and an alternative for people (like himself) who were allergic to cow's milk.

Definition

Evaporated milk is fresh, homogenized milk from which 60 percent of the water has been removed. After the water has been removed, the product is chilled, stabilized, packaged and sterilized. A slightly caramelized flavour results from the high heat process, and it is slightly darker in colour than fresh milk. The evaporation process also concentrates the nutrients and the food energy. Thus, for the same weight, undiluted evaporated milk contains more food energy than fresh milk.

In Malaysia

In Malaysia, due to price controls, evaporated (and condensed) milk contains palm oil. It is one of the ingredients to make *Teh Tarik* in Malaysia and Singapore.

Notable Producers

Evaporated milk is sold by several manufacturers:

*   Carnation Evaporated Milk (the brand is now owned by Nestle and licensed to Smuckers in Canada)
*   PET Evaporated Milk (now owned by Smuckers)
*   Magnolia evaporated milk-(now produced by Eagle Family Foods)
*   Nestlé evaporated milk
*   F&N Evaporated Milk.

Scalded Milk

Scalded milk is milk that has been heated to 82°C/180°F. At this temperature, bacteria and enzymes in the milk are destroyed. Since most milk sold today is pasteurized, which accomplishes both of these goals, milk is typically scalded simply to increase its temperature. During scalding, a cooking utensil, a milk watcher, is used to prevent both boiling over and scorching(burning) of the milk.

Uses

*   Scalded milk is called for in the original recipes for béchamel sauce, to prevent the sauce from thickening excessively. Since these early recipes predate pasteurization, this was a necessary step.
*   Scalded milk is used in bread to make a more tender loaf.
*   Scalded milk is used in yogurt to make the proteins unfold.

The acid produced during the yogurt development causes less whey separation and a firmer yogurt.

- Café au lait, baked milk, and ryazhenka also use scalded milk.
- Scalded milk is used in many doughnut recipes.

Powdered Milk

Powdered milk is a manufactured dairy product made by evaporating milk to dryness. One purpose of drying milk is to preserve it; milk powder has a far longer shelf life than liquid milk and does not need to be refrigerated, due to its low moisture content. Another purpose is to reduce its bulk for economy of transportation. Powdered milk and dairy products include such items as dry whole milk, non-fat dry milk, dry buttermilk, dry whey products and dry dairy blends.

History and Manufacture

While Marco Polo wrote of Mongolian Tatar troops in the time of Kublai Kahn carrying sun-dried skimmed milk as "a kind of paste", the first usable commercial production of dried milk was invented by the Russian chemist M. Dirchoff in 1832. In 1855 T.S. Grimwade took a patent on a dried milk procedure, though a William Newton had patented a vacuum drying process as early as 1837. Today, powdered milk is usually made by spray drying nonfat skim milk, whole milk, buttermilk or whey. Pasteurized milk is first concentrated in an evaporator to about 50% milk solids. The resulting concentrated milk is sprayed into a heated chamber where the water almost instantly evaporates, leaving fine particles of powdered milk solids.

Alternatively, the milk can be dried by drum drying. Milk is applied as a thin film to the surface of a heated drum, and the dried milk solids are then scraped off. Powdered milk made this way tends to have a cooked flavour, due to caramelization caused by greater heat exposure.

Another process is freeze drying, which preserves many nutrients in milk, compared to drum drying.

The drying method and the heat treatment of the milk as it is processed alters the properties of the milk powder (for example, solubility in cold water, flavour, bulk density).

Uses

Powdered milk is frequently used in the manufacture of infant formula, confectionery such as chocolate and caramel candy, and in recipes for baked goods where adding liquid milk would render the product too thin. Powdered milk is also widely used in various sweets such as the famous Indian milk balls known as Rasa-Gulla and popular Pakistani sweet delicacy (sprinkled with desiccated coconut) known as Chum chum (made with skim milk powder).

Powdered milk is also a common item in UN food aid supplies, fallout shelters, warehouses, and wherever fresh milk is not a viable option. It is widely used in many developing countries because of reduced transport and storage costs (reduced bulk and weight, no refrigerated vehicles).

As with other dry foods, it is considered nonperishable, and is favoured by survivalists, hikers, and others requiring nonperishable, easy-to-prepare food.

Reconstituting one cup of milk from powdered milk requires one cup of potable water and one-third cup of powdered milk. Powdered milk is also used in western blots as a blocking buffer to prevent nonspecific protein interactions, and is referred to as Blotto.

Food and Health

Nutritional Value: Milk powders contain all twenty standard amino acids (the building blocks of proteins) and are high in soluble vitamins and minerals. According to USAID the typical average amounts of major nutrients in the unreconstituted milk are (by weight) 36% protein, 52% carbohydrates (predominantly lactose), calcium 1.3%, potassium 1.8%. Their milk powder is fortified with Vitamin A and D, 3000IU and 600IU respectively per 100g. Inappropriate storage conditions (high relative humidity and high ambient temperature) can significantly degrade the nutritive value of milk powder. Commercial milk powders are reported to contain oxysterols (oxidized cholesterol) in higher amounts than in fresh milk (up to 30ìg/g, versus trace amounts in fresh milk). The oxysterol free radicals have been suspected of being initiators of atherosclerotic plaques. For comparison, powdered eggs contain even more oxysterols, up to 200ìg/g.

Market

Fonterra, a New Zealand based multinational company, is the world's largest producer of milk powder controlling 40 percent of the global whole milkpowder. The dominance of New Zealand in the global dairy industry, for example Fonterra controls around 30% of the world's dairy exports, has prompted the formation of a futures market for trading whole milkpowder.

Adulteration

In the 2008 Chinese milk scandal, melamine adulterant was found in Sanlu infant formula, added to fool tests into reporting higher protein content. Thousands became ill and some children died after consuming the product.

Dried Milk or Milk Powder

Roller Dried Milk Powder

Dried milk (Milk Powder) products play a significant role in conserving the milk solids since their biological value can be retained for a long period of time under relatively simple storage conditions. The roller dried powder finds special application in the manufacture of infant foods, confectionery, ice cream, milk sweets etc. Milk is concentrated by means of condensation to about 20-25 per cent of total solids in a vacuum pan or evaporator and fed to continuously revolving drum drier, which are internally heated with steam. The film of the dried product is continuously scrapped off by a stationary knife or doctor blade, located opposite in the point of application of milk. The dried milk film thus obtained is ground, which is then sifted, packed and stored.

Method of Manufacture of Dried Milk

- The milk is separated (in case of skim milk powder) or clarified and standardized (in case of whole milk powder).
- The milk is fore-warmed to 95°C and re-condensed to 20% total solids. Homogenization of the condensed milk is done in case of manufacture of whole milk powder.
- The roller drums are observed for its correct alignment with the knife, cleanliness of the surface of the rollers, pressure build up in the drum and working condition of the powder conveyors and exhaust fan.
- The condensed milk in the feed line is taken in to the feed trough and the scraping blades are adjusted. The rate of milk flow is ensured to the determined level in the trough and the dried milk is scraped off from the surface of the drum uniformly.
- The semi dried and charred powders obtained initially have to be discarded. The conveyors are adjusted properly to carry the powder.
- When the milk flow is over, the conveyors are stopped and some hot water is run over the drums.

- The knife is taken off from the surface of the drum and supply of steam to the drum is cut.
- The rollers are stopped only when there is no steam pressure inside and the rollers are sufficiently cool.
- The vapour exhaust fan is stopped.

Spray Dried Milk Powder

Spray drying is the most important method of drying milk and milk products. By spraying in to a stream of hot air, the droplets formed present an extremely large amount of surface area and get dried immediately due to rapid evaporation of moisture. Milk is preheated and concentrated to 40-45% percent total solids. Hot air is filtered and directed in to the drying chamber. The concentrate is atomized to obtain small particles ranging from 10-100m in diameter. The air leaving the drying chamber enters a cyclone separator where the fine particles are collected. The dried products are cooled, sifted and packed in suitable packaging material.

Method of Manufacture of Spray Dried Milk Powder

- The milk is separated (in case of skim milk powder) or clarified and standardized (in case of whole milk powder).
- The milk is fore-warmed to 95°C and re-condensed to 40% total solids. Homogenization of the condensed milk is done in case of manufacture of whole milk powder.
- The spray drier is cleaned, pipelines are connected and general conditions of the valves and cyclone separators are checked.
- The steam is let in to the air heating coils. The exhaust fan is switched on, keeping the main door partially open.
- The blower is started and the main door is closed.
- The temperature of the hot air is adjusted to 180°C and this temperature is maintained for ten minutes to ensure the sterility of the unit.
- The atomizer is started and allowed to obtain the required speed.
- Initially, 20 litres of water is used for spraying through the feed pump.
- When the outlet air temperature reaches 100°C, the feed is changed to concentrate. The temperature of the inlet air entering the drying chamber usually ranges from 150-260°C

and the outlet air exhausted from the drying chamber ranges from 100-105°C. the relative humidity of the drying chamber is quite low i.e. 3-4%.

- The *milk powder* is collected in the powder silo.

- When the flow of milk is over, the balance tank is flushed with about 20 litres of hot water (90°C) and the flow of milk to the dryer is reduced immediately.

- The atomizer and feed pump are stopped when the feed is empty.

- The steam supply to the air heating coils is stopped.

- The exhaust fan and air blower are stopped.

- The dried milk or milk powder is packed in required quantity in suitable containers.

Chapter 5

Biogas Technology

Biogas typically refers to a gas produced by the biological breakdown of organic matter in the absence of oxygen. Biogas originates from biogenic material and is a type of biofuel. Biogas is produced by anaerobic digestion or fermentation of biodegradable materials such as biomass, manure, sewage, municipal waste, green waste, plant material and energy crops. This type of biogas comprises primarily methane and carbon dioxide. Other types of gas generated by use of biomass is wood gas, which is created by gasification of wood or other biomass. This type of gas consist primarily of nitrogen, hydrogen, and carbon monoxide, with trace amounts of methane. The gases methane, hydrogen and carbon monoxide can be combusted or oxidized with oxygen. Air contains 21% oxygen. This energy release allows biogas to be used as a fuel. Biogas can be used as a low-cost fuel in any country for any heating purpose, such as cooking. It can also be used in modern waste management facilities where it can be used to run any type of heat engine, to generate either mechanical or electrical power. Biogas can be compressed, much like natural gas, and used to power motor vehicles and in the UK for example is estimated to have the potential to replace around 17% of vehicle fuel. Biogas is a renewable fuel, so it qualifies for renewable energy subsidies in some parts of the world.

Ancient Persians observed that rotting vegetables produce flammable gas. In the 13th century, the traveller Marco Polo noted the Chinese used covered sewage tanks to generate power, while biogas technologies were also referred to by 17th century author Daniel Defoe.

In 1859, an anaerobic digestion plant was built to process sewage at a Bombay leper colony. Biogas has been used in the UK since 1895, when gas from sewage was used in street lamps across the city of Exeter.

Production

Biogas is practically produced as landfill gas (LFG) or digester gas.

A biogas plant is the name often given to an anaerobic digester that treats farm wastes or energy crops.

Biogas can be produced utilizing anaerobic digesters. These plants can be fed with energy crops such as maize silage or biodegradable wastes including sewage sludge and food waste. During the process, an air-tight tank transforms biomass waste into methane producing renewable energy that can be used for heating, electricity, and many other operations that use any variation of an internal combustion engine, such as GE Jenbacher gas engines. There are two key processes: Mesophilic and Thermophilic digestion.

Landfill gas is produced by wet organic waste decomposing under anaerobic conditions in a landfill. The waste is covered and mechanically compressed by the weight of the material that is deposited from above. This material prevents oxygen exposure thus allowing anaerobic microbes to thrive. This gas builds up and is slowly released into the atmosphere if the landfill site has not been engineered to capture the gas. Landfill gas is hazardous for three key reasons. Landfill gas becomes explosive when it escapes from the landfill and mixes with oxygen. The lower explosive limit is 5% methane and the upper explosive limit is 15% methane. The methane contained within biogas is 20 times more potent as a greenhouse gas than carbon dioxide. Therefore uncontained landfill gas which escapes into the atmosphere may significantly contribute to the effects of global warming. In addition landfill gas' impact in global warming, volatile organic compounds (VOCs) contained within landfill gas contribute to the formation of photochemical smog.

Composition

Table : Typical composition of biogas

Compound	Chem	%
Methane	CH_4	50–75
Carbon dioxide	CO_2	25–50
Nitrogen	N_2	0–10
Hydrogen	H_2	0–1
Hydrogen sulfide	H_2S	0–3
Oxygen	O_0	0–0

The composition of biogas varies depending upon the origin of the anaerobic digestion process. Landfill gas typically has methane concentrations around 50%. Advanced waste treatment technologies can produce biogas with 55–75% CH_4 or higher using in situ purification techniques As-produced, biogas also contains water vapour, with the fractional water vapour volume a function of biogas temperature; correction of measured volume for water vapour content and thermal expansion is easily done via algorithm.

In some cases biogas contains siloxanes. These siloxanes are formed from the anaerobic decomposition of materials commonly found in soaps and detergents. During combustion of biogas containing siloxanes, silicon is released and can combine with free oxygen or various other elements in the combustion gas. Deposits are formed containing mostly silica (SiO_2) or silicates (Si_xO_y) and can also contain calcium, sulfur, zinc, phosphorus. Such white mineral deposits accumulate to a surface thickness of several millimeters and must be removed by chemical or mechanical means.

Practical and cost-effective technologies to remove siloxanes and other biogas contaminants are currently available.

Applications

Biogas can be utilized for electricity production on sewage works, in a CHP gas engine, where the waste heat from the engine is conveniently used for heating the digester; cooking; space heating; water heating; and process heating. If compressed, it can replace compressed natural gas for use in vehicles, where it can fuel an internal combustion engine or fuel cells and is a much more effective displacer of carbon dioxide than the normal use in on-site CHP plants.

Methane within biogas can be concentrated via a biogas upgrader to the same standards as fossil natural gas(which itself has had to go through a cleaning process), and becomes biomethane. If the local gas network allows for this, the producer of the biogas may utilize the local gas distribution networks.

Gas must be very clean to reach pipeline quality, and must be of the correct composition for the local distribution network to accept. Carbon dioxide, water, hydrogen sulfide and particulates must be removed if present. If concentrated and compressed it can also be used in vehicle transportation. Compressed biogas is becoming widely used in Sweden, Switzerland, and Germany. A biogas-powered train has been in service in Sweden since 2005.

Biogas has also powered automobiles. In 1974, a British documentary film entitled *Sweet as a Nut* detailed the biogas production process from pig manure, and how the biogas fueled a custom-adapted combustion engine.

Benefits

By using biogas, many advantages arise. In North America, utilization of biogas would generate enough electricity to meet up to three percent of the continent's electricity expenditure. In addition, biogas could potentially help reduce global climate change. Normally, manure that is left to decompose releases two main gases that cause global climate change: nitrous dioxide and methane. Nitrous dioxide warms the atmosphere 310 times more than carbon dioxide and methane 21 times more than carbon dioxide. By converting cow manure into methane biogas via anaerobic digestion, the millions of cows in the United States would be able to produce one hundred billion kilowatt hours of electricity, enough to power millions of homes across the United States. In fact, one cow can produce enough manure in one day to generate three kilowatt hours of electricity; only 2.4 kilowatt hours of electricity are needed to power a single one hundred watt light bulb for one day. Furthermore, by converting cow manure into methane biogas instead of letting it decompose, we would be able to reduce global warming gases by ninety-nine million metric tons or four percent.

The 30 million rural households in China that have biogas digesters enjoy 12 benefits: saving fossil fuels, saving time collecting firewood, protecting forests, using crop residues for animal fodder instead of fuel, saving money, saving cooking time, improving hygienic conditions, producing high-quality fertilizer, enabling local mechanization and electricity production, improving the rural standard of living, and reducing air and water pollution.

Biogas Upgrading

Raw biogas produced from digestion is roughly 60% methane and 29% CO_2 with trace elements of H_2S, and is not high quality enough if the owner was planning on selling this gas or using it as fuel gas for machinery. The corrosive nature of H_2S alone is enough to destroy the internals of an expensive plant. The solution is the use of a biogas upgrading or purification process whereby contaminants in the raw biogas stream are absorbed or scrubbed, leaving 98% methane per unit volume of gas. There are four main methods of biogas upgrading, these include water washing, pressure swing absorption, selexol absorption and chemical treatment. The most prevalent method is water washing where high pressure gas flows into a column where the carbon dioxide and other trace elements are scrubbed by cascading water running counter-flow to the gas. This arrangement can deliver 98% methane with manufacturers guaranteeing maximum 2% methane loss in the system. It takes roughly between 3-6% of the total energy output in gas to run a biogas upgrading system.

Biogas Gas-grid Injection

Gas-grid injection is the injection of biogas into the methane grid (natural gas grid). Injections includes biogas: until the breakthrough of micro combined heat and power two-thirds of all the energy produced by biogas power plants was lost (the heat), using the grid to transport the gas to customers, the electricity and the heat can be used for on-site generation resulting in a reduction of losses in the transportation of energy. Typical energy losses in natural gas transmission systems range from 1–2%. The current energy losses on a large electrical system range from 5–8%.

In the field of renewable energy, biogas technology refers to systems designed to turn organic waste products into usable energy. Biogas is a kind of gas that is produced during the anaerobic processing of organic matter such as manure, plant matter, or even municipal waste materials. Biogas typically consists mainly of methane, with a significant proportion of carbon dioxide, and smaller quantities of other gases such as nitrogen and hydrogen.

Biogas fuel is a flammable substance that burns in a similar fashion to liquefied petroleum gas (LPG), and as such, biogas energy can be utilized as an alternative to fossil fuels.

Biogas production is often achieved using a biogas plant, which is a system that "digests" organic matter to produce gas. Biogas technology is often used on farms with the primary goal of controlling waste pollution. Dairy farms, for example, may have significant problems with manure polluting water sources. Biogas technology allows farmers to address this problem with the beneficial side effect of also creating a new power source.

Biogas electricity and energy for heating can be produced locally with biogas technology systems on some farms. For example, a farmer with 500 cows could install a system that would process the manure produced by the cows, and could generate sufficient electricity to power the entire dairy system. Larger scale systems in some countries collect manure from multiple farms, and perform the digestion process centrally in a large plant.

Similar biogas technology may be used in sewage treatment. Wastewater contains organic matter, including organic solids, and anaerobic digestion may be used to break down these solids. This process, often called sludge digestion, may also reduce the levels of harmful bacteria in the water.

There are three main types of biogas technology digesters: covered lagoon, complete mix, and plug-flow. All of these types digest manure to produce biogas. They differ in their efficiency, the concentration of solids they can handle, and their suitability to different climates and applications.

Covered lagoon digesters are the simplest type, consisting of a storage pool to hold the manure, and a cover to trap the gas. This type of technology is suitable for warm climates, and is mainly used for liquid manure with a small percentage of solid matter. Complete mix digesters hold manure in a heated tank, and have a mechanical mixing device to speed up the digestion process. Plug-flow digesters are more complex biogas technology systems that actively pass the manure being processed through the system, as it is digested.

- Dairy farms, for example, may have significant problems with manure polluting water sources. Biogas technology allows farmers to address this problem with the beneficial side effect of also creating a new power source.

- Unintentional production of biogases has been an ongoing issue in many regions of the world, and several nations have also looked to biogas as a potential source of clean energy. India and China have both invested extensively in creative biogas technology to provide fuel for their citizens and there are a number of interesting applications for this gas which appeal to people who are interested in sustainable energy and the health of the environment.

Biogas Plants for Organic Farming

The biogas plant consists of two components: a digester (or fermentation tank) and a gas holder. The digester is a cube-shaped or cylindrical waterproof container with an inlet into which the fermentable mixture is introduced in the form of a liquid slurry. The gas holder is normally an airproof steel container that, by floating like a ball on the fermentation mix, cuts off air to the digester (anaerobiosis) and collects the gas generated. In one of the most widely used designs, the gas holder is equipped with a gas outlet, while the digester is provided with an overflow pipe to lead the sludge out into a drainage pit.

Singh, and others (1, 3) have documented several guidelines for consideration in the designing of batch (periodic feeding) and continuous (daily feeding) compartmentalized and non-compartmentalized biogas plants that are of either the vertical or horizontal type. In addition, Loll (18) has recently dealt with the scientific principles, process engineering, and shapes of digestion reactors, and with the economics of the technology.

Digester reactors are constucted from brick, cement, concrete, and steel. In Indonesia, where rural skills in brick making, brick laying, plastering, and bamboo craft are well established, clay bricks have successfully replaced cement blocks and concrete. In areas where the cost is high, the "sausage" or bag digester (14) appears to be ideal. The digester is constructed of 0.55 mm thick Hypalon laminated with Neoprene and reinforced with nylon. The bag is fitted with an inlet and an outlet made from PVC. Even if imported from the United States, the cost of the digester and the gas holder (both combined in one bag) is only 10 per cent of that for a concrete-steel digester. Another advantage is that it can be mass produced and is easily mailed. In rural areas, the whole installation is completed in a matter of minutes. A hole in the ground accommodates the bag, which is filled two-thirds full with waste water. Gas production fully inflates the bag, which is weighted down and fitted with a compressor to increase gas pressure.

Environmental and Operational Considerations

Raw Materials (19)

Raw materials may be obtained from a variety of sources- livestock and poultry wastes, night soil, crop residues, food-processing and paper wastes, and materials such as aquatic weeds, water hyacinth, filamentous algae, and seaweed. Different problems are encountered with each of these wastes with regard to collection, transportation, processing, storage, residue utilization, and ultimate use. Residues from the agricultural sector such as spent straw, hay, cane trash, corn and plant stubble, and bagasse need to be shredded in order to facilitate their flow into the digester reactor as well as to increase the efficiency of bacterial action. Succulent plant material yields more gas than dried matter does, and hence materials like brush and weeds need semi-drying. The storage of raw materials in a damp, confined space for over ten days initiates anaerobic bacterial action that, though causing some gas loss, reduces the time for the digester to become operational.

Influent Solids Content (16, 19, 21)

Production of biogas is inefficient if fermentation materials are too dilute or too concentrated, resulting in, low biogas production and insufficient fermentation activity, respectively. Experience has shown that the raw-material (domestic and poultry wastes and manure) ratio to water should be 1:1, i.e., 100 kg of excrete to 100 kg of water. In the slurry, this corresponds to a total solids concentration of 8- 11 per cent by weight.

Loading (14, 19)

The size of the digester depends upon the loading, which is determined by the influent solids content, retention time, and the digester temperature. Optimum loading rates vary with different digesters and their sites of location. Higher loading rates have been used when the ambient temperature is high. In general, the literature is filled with a variety of conflicting loading rates. In practice, the loading rate should be an expression of either (a) the weight of total volatile solids (TVS) added per day per unit volume of the digester, or (b) the weight of TVS added per day per unit weight of TVS in the digester. The latter principle is normally used for smooth operation of the digester.

Seeding (14, 19)

Common practice involves seeding with an adequate population of both the acid-forming and methanogenic bacteria. Actively digesting sludge from a sewage plant constitutes ideal "seed" material. As a general guideline, the seed material should be twice the volume of the fresh manure slurry during the start-up phase, with a gradual decrease in amount added over a three-week period. If the digester accumulates volatile acids as a result of overloading, the situation can be remedied by reseeding, or by the addition of lime or other alkali.

pH (14, 19)

Low pH inhibits the growth of the methanogenic bacteria and gas generation and is often the result of overloading. A successful pH range for anaerobic digestion is 6.0- 8.0; efficient digestion occurs at a pH near neutrality. A slightly alkaline state is an indication that pH fluctuations are not too drastic. Low pH may be remedied by dilution or by the addition of lime.

Temperature (13,14,19, 21)

With a mesophilic flora, digestion proceeds best at 30- 40 C; with thermophiles, the optimum range is 50- 60 C. The choice of the temperature to be used is influenced by climatic considerations In general, there is no rule of thumb, but for optimum process stability, the temperature should be carefully regulated within a narrow range of the operating temperature. In warm climates, with no freezing temperatures, digesters may be operated without added heat. As a safety measure, it is common practice either to bury the digesters in the ground on account of the advantageous insulating properties of the soil, or to use a greenhouse covering. Heating requirements and, consequently, costs, can be minimized through the use of natural materials such as leaves, sawdust, straw, etc., which are composted in batches in a separate compartment around the digester.

Nutrients (13,17,19, 21)

The maintenance of optimum microbiological activity in the digester is crucial to gas generation and consequently is related to nutrient availability. Two of the most important nutrients are carbon and nitrogen and a critical factor for raw material choice is the overall C/N ratio.

Domestic sewage and animal and poultry wastes are examples of N-rich materials that provide nutrients for the growth and multiplication of the anaerobic organisms. On the other hand, N-poor materials like green grass, corn stubble, etc., are rich in carbohydrate substances that are essential for gas production. Excess availability of nitrogen leads to the formation of NH3, the concentration of which inhibits further growth. Ammonia toxicity can be remedied by low loading or by dilution. In practice, it is important to maintain, by weight, a C/N ratio close to 30:1 for achieving an optimum rate of digestion. The C/N ratio can be judiciously manipulated by combining materials low in carbon with those that are high in nitrogen, and vice versa.

Toxic Materials (13,14,19)

Wastes and biodegradable residue are often accompanied by a variety of pollutants that could inhibit anaerobic digestion. Potential toxicity due to ammonia can be corrected by remedying the C/N ratio of manure through the addition of shredded bagasse or straw, or by dilution. Common toxic substances are the soluble salts of copper, zinc, nickel, mercury, and chromium. On the other hand, salts of sodium, potassium, calcium, and magnesium may be stimulatory or toxic in action, both manifestations being associated with the cation rather than the anionic portion of the salt. Pesticides and synthetic detergents may also be troublesome to the process.

Stirring (13,14,17- 19, 21)

When solid materials not well shredded are present in the digester, gas generation may be impeded by the formation of a scum that is comprised of these low-density solids that are enmeshed in a filamentous matrix. In time the scum hardens, disrupting the digestion process and causing stratification. Agitation can be done either mechanically with a plunger or by means of rotational spraying of fresh influent. Agitation, normally required for bath digesters, ensures exposure of new surfaces to bacterial action, prevents viscid stratification and slow-down of bacterial activity, and promotes uniform dispersion of the influent materials throughout the fermentation liquor, thereby accelerating digestion.

Retention Time (19, 21)

Other factors such as temperature, dilution, loading rate, etc., influence retention time. At high temperature bio-digestion occurs faster, reducing the time requirement. A normal period for the digestion of dung would be two to four weeks.

Developments and Processes for Rural Areas

Two years ago, the Economic and Social Council of the United Nations adopted a survey, presented in 1978 to the Committee on Science and Technology for Development, listing the on-going research and development in unconventional sources of energy. From the point of view of the developing countries, it is heartening to note that the "use of farm wastes to produce methane" has also been identified in the United Nations World Plan of Action for the Application of Science and Technology to Development.

The Economic and Social Council for Asia and the Pacific, moreover, adopted the Colombo Declaration at its thirtieth session, which determined that the most urgent priorities for action are in the fields of food, energy, raw materials, and fertilizers, and that these priorities would be best met by the integrated biogas system (IBS). An integrated system aims at the facile generation of fertilizer and acquisition of energy, production of protein via the growth of algae and fish in oxidation ponds, hygienic disposal of sewage and other refuse, and is a tangible effort to counteract environmental pollution. The heart of the system is the biogas process; it has the potential to "seed" self-reliance in relatively primitive economies (14, 22, 23). Allied benefits include the development of rural industry, the provision of local job opportunities, and the progressive eradication of hunger and poverty.

The coupling of a photosynthetic step (24- 26) with digestion provides for the transformation of the minerals left by digestion directly into algae that can then be used as fodder, as feed for fish, as fertilizer, or for increased energy production by returning them to the digester process.

The IBS aims at putting back into soil and water what has been taken from them, and increasing the amounts of nutrients by fixing CO2 and N2 from the atmosphere into the soil and water through photosynthesis by algae. Involving low cash investments on a decentralized basis, the implementation of IBS provides employment to the whole work force without disruption of the rural structure. Furthermore, it is an apt example of soft technology that does not pollute or destroy the physical environment. At the College of Agriculture of the University of the Philippines, preliminary work on a small scale has begun. In England, an Eco-house has been built by Graham Caine on the Thames Polytechnical Playing Fields at Eltham, southeast of London. Results on the project, however, are not yet available.

Cost-benefit Analyses

There is no general answer to the economic feasibility of biogas production. National economic considerations play an important role. In Korea, wood is in short supply (27) and domestic fuel substitutes like rice and barley straw, and coal and oil could be conserved; wood could be a foreign-exchange earner in the field of handicrafts. In India, transportation costs of coal and oil to the rural areas is high and an extra burden on an already poor farmer.

The rural share in the energy consumption of electricity and coal is not considerable because, as the Report of the Panel of the National Committee of Science and Technology on Fuel and Power indicates, the large towns and cities with populations of 500,000 and more accommodate only 6 per cent of India's total population but consume about 50 per cent of the total commercial energy produced in the country.

In the villages, however, kerosene is used for lighting, but it is clear that with increasing population, biogas generation seems to offer solutions in the areas of fuel availability, electricity, fertilizer for cash crops, and would provide other socio-economic benefits.

On the other hand, cost-benefit analyses of methane generation vary widely, depending upon the uses and actual benefits of biogas production, public and private costs associated with the development and utilization of methane, and on the technology used to generate methane. Several factors have been listed in the economics of biogas generation (14, 17- 19, 28). An appropriate example is the fact that a village-model gas plant, which cost Rs 500 some years ago, cost Rs 1,500 in 1974 and Rs 2,000 in 1977. Hence, a significant problem is whether rural people who cannot spend Rs 2,000 can cope with increasing inflationary and digester construction material costs.

The Khadi and Village Industries Commission has helped to tackle the problem through rural community co-operation and a scheme of subsidies and loans to encourage individual families, groups of families, institutions, and communities to construct biogas plants. The net annual income of approximately US$60 shows that the capital investment of US$340 can be recouped in about six years. There are also incidental advantages of hygienic improvement, the absence of smoke and soot in gas burning, convenience in burning, and the increased richness of manure.

Health Hazards

Health hazards are associated with the handling of night soil and with the use of sludge from untreated human excrete as fertilizer. In general, published data indicate that a digestion time of 14 days at 35 C is effective in killing (99.9 per cent die-off rate) the enteric bacterial pathogens and the enteric group of viruses.

However, the die-off rate for roundworm (Ascaris lumbricoides) and hookworm (Ancylostoma) is only 90 per cent, which is still high. In this context, biogas production would provide a public health benefit beyond that of any other treatment in managing the rural health environment of developing countries.

Production Technology of Paddy Straw Mushroom

The steps involved in cultivation are; choice of substratum, bed-preparation and cropping, after care of beds, and harvesting and marketing.

Choice of Substratum

Paddy straw is considered to be the best; other which can be used are wheat straw, jowar, maize, rye straw, sugarcane by gasse, tobacco, and banana leaves.

Bed-preparation and Cropping

Firstly, hand-harvested, 3-4' long. Well dried and disease-free paddy straw are taken to prepare their bundles. 35 bundles of paddy straw are required for one bed and each bundle should be 1 to 11/2 kg in weight. The bundles are soaked in water for 8-16 hours, are taken out of water, washed with fresh water and allowed to drain off excess water.

Now, a bed is prepared by putting four layers of the paddy straw bundles one over the other each layer contains 8 bundles. The spawn bottle is opened and raked with a woolen or glass rod.

The spawn is now sprinkled by hand all over on the margin of the bed about 10 cms. Away from the edge and continuing upto 23 cms. Inside. Thus, the central portion has been left spawning in 1st layer. The sprinkled spawn is covered with a light dusting of 'besan' (gram powder). The second, third and fourth layers are also prepared in the same way as in case of 1st but the difference in 4th layer is that the sprinkling of spawn is done on the entire surface instead of periphery as in case of rests of the layers. The layer should also be dusted by the gram powder and must be covered with a thin layer of straw upto 8 cms. Thickness. This is also done with the remaining three bundles of straw.

After Care of Beds

Water is sprayed 2-3 times in hot day and 1-2 times in rainy season. If necessary, 0.1% Malathion and 0.2% Dithane Z-78 is sprayed to overcome insects, pests, and other diseases.

Harvesting and Marketing

Cropping starts 10-12 days after spawning and remains upto 15-20 days. Mushroom is harvested at button stage or just after rupturing of the cup (pileus). The harvesting is done by twisting so that broken pieces are not left in the beds otherwise bacterial rotting generally starts and spreads to other healthy mushrooms. After harvesting the mushrooms should be used within 8 hours or kept 10-15⁰C for 24 hours otherwise they get spoiled. One may keep them for a week in refrigerator.

Fresh mushrooms are dried either in sun or in even at 55-60⁰C for 8 hours. After drying, they are packed and sealed otherwise they absorb moisture which spoils them. The packing of mushrooms is preferred only in button stage. However, one can find 3-4 kg yield/ bed of mushrooms.

Microbial Process in Mushroom Production

The mushroom is a fungus and is quite finicky about its food source. Mushrooms lack the ability to use energy from the sun. They are not green plants because they do not have chlorophyll. Mushrooms extract their carbohydrates and proteins from a rich medium of decaying, organicmatter vegetation. This rich organic matter must be prepared into nutrient- rich substrate composts that the mushroom can consume. When correctly made, this food may become available exclusively to the mushroom and would not support the growth of much else. At a certain stage in the decomposition, the mushroom grower stops the process and plants the mushroom so it becomes the dominant organism in that environment.

The sequence used to produce this specific substrate for the mushroom is called composting or compost substrate preparation and is divided into two stages, Phase I and Phase II. Each stage has distinct goals or objectives. It is the grower's responsibility to provide the necessary ingredients and environmental conditions for the chemical and biological processes required to complete these goals. The management of starting ingredients and the proper conditions for composting make growing mushrooms so demanding.

Making a Composted Substrate

Many agricultural by-products are used to make mushroom substrate. Straw-bedded horse manure and hay or wheat straw are the common bulk ingredients. "Synthetic" composts are those in which the prime ingredient is not straw-bedded horse manure. If bulk ingredients are high in nitrogen, other high-carbohydrate bulk ingredients—such as corncobs, cottonseed hulls, or cocoa bean hulls— are added to the mix. All compost formulas require the addition of nitrogen supplements and gypsum. Additional nitrogen-rich supplements are added to composts to increase the nitrogen content to 1.5–1.7 percent for horse manure or 1.7–1.9 percent for synthetic;

both are computed on a dry weight basis. Poultry manure is probably the most common and economical source of nitrogen. A variety of meals or seeds, such as cottonseed meal, soybean meal, or brewer's grain may also be used. Inorganic or nonprotein nitrogen sources such as ammonia nitrate and urea are also used, but only in small amounts when high-carbohydrate bulk ingredients are used. Gypsum is added to minimize "greasiness" and to buffer the pH of the compost.

Gypsum increases the flocculation of colloids in the compost, which prevents the straws from sticking together and inhibiting air penetration.

Air, which supplies oxygen to the microbes and chemical reactions, is essential to the composting process. Gypsum may be added early in the composting process, at 70–100 lbs per ton of dry ingredients.

A concrete slab, referred to as a wharf, is required for composting. In addition, a compost turner to aerate and water the ingredients and a tractor-loader to move the ingredients to the turner are needed. Water used during a substrate preparation operation can be recycled back into the process. It is, in a sense, a closed system. Water runoff into the environment is nonexistent on a properly managed substrate preparation wharf. Water collected in concrete pits or a sealed lagoon is aerated and recycled to soak bulk ingredients before the composting process begins.

Conventional Phase I composting begins by mixing and wetting the ingredients as they are stacked. Most farms have a preconditioning phase in which bulk ingredients and some supplements are watered and stacked in a large pile for several days to soften, making them more receptive to water. This preconditioning time may range from 3 to 15 days. The piles are turned daily or every other day. After this pre-wet stage, the compost is formed into a rectangular pile with tight sides and a loose centre.

A compost turner is typically used to form this pile. Water is sprayed onto the horse manure or synthetic compost as these materials move through the turner. Nitrogen supplements and gypsum can be spread over the top of the bulk ingredients and are thoroughly mixed by the turner.

Once the pile is wetted and formed, aerobic fermentation (composting) commences as microbial growth and reproduction naturally occur in the bulk ngredients. Heat, ammonia, and carbon dioxide (CO_2) are released as by-products during this process. Compost activators, other than those mentioned, are not needed.

As temperatures increase above 155°F (70°C), microorganisms cease growing and a chemical reaction begins.

Concentrating and preserving complex carbohydrates is one goal of Phase I. The quantity and the quality of nitrogen in the system are changed to a type of nitrogen that Phase II microorganisms and, eventually, the mushroom will use as food.

Adequate moisture, oxygen, nitrogen, and carbohydrates must be present throughout the process; otherwise, the process will stop. This is why water and supplements are added periodically and the compost pile is aerated as it moves through the turner. Oxygenation is achieved in conventional outdoor ricks by natural convection.

The high pile temperatures draw ambient air through the sides of the stack, and as the air is heated, it rises upward through the stack—a process commonly referred to as the chimney effect. The sides of the pile should be firm and dense, yet the centre must remain loose throughout Phase I composting. The exclusion of air results in an airless (anaerobic) environment. As the straw or hay softens during composting, the materials become less rigid and more compact while substrate density increases. Thus, less air reaches the bottom and centre of the pile. A lack of oxygen may occur after the large quantities of water are added to the dry bulk ingredients and before sufficient heat is generated to start the draw of air into the pile. Under anaerobic conditions, organic acids and other deleterious chemical compounds are formed. Therefore, preparing substrate under aerobic conditions, where less offensive odors are produced, is better for mushroom growers.

Aerated Phase I Composting

Improving community relations has led to alterations in the way the Phase I mushroom composting process is carried out. As urban areas encroach on rural farmland, residents have made odor-related complaints and legal battles have ensued, which suggest a need for more stringent odor-management practices.

If the pile is not turned and aerated during Phase I composting, oxygen may become limited and anaerobic conditions may develop along the bottom of the stack. As the anaerobic core gets larger, more offensive odors are produced. In order to maintain aerobic conditions throughout the entire substrate pile, supplemental aeration is sometimes used. This aeration is accomplished by using a fan to force air up through a concrete pad with a series of evenly distributed openings and into the substrate material. This design is referred to as an aerated floor. Systems have been built with structural sidewalls, usually of concrete and occasionally of wood, to form the piles with a uniform height and depth. Aside from aerated floors and structural sidewalls, there is great variation among bunker systems currently being used for Phase I.

Aerated composting systems are replacing conventional ricks throughout Europe and are beginning to gain acceptance in North America as the quest to manage odors continues.

Europeans were the first to regulate emissions from their agricultural operations. Therefore, most European mushroom composting operations have employed some type of enclosed or environmentally controlled Phase I system. In North America, a few systems have been built to test the technology. Eventually they may become common at commercial operations. Unfortunately, little information is available to show how these systems reduce emissions.

Therefore, determining how effective aerated systems are in reducing odors is difficult.

Phase I is considered complete as soon as the raw ingredients become pliable and are capable of holding water, the odor of ammonia is sharp, and the dark-brown colour indicates that carmelization and browning reactions have occurred. At the beginning of Phase I, the substrate is bulky and yellow. At the end of Phase I substrate preparation, the substrate should be dense, chocolate brown in colour, and have a strong odor of ammonia. The substrate still has some structure so aeration can be maintained during Phase II composting. The potential fresh mushroom yield depends on the amount of dry weight filled. In order to achieve a substrate density in the growing structure necessary to support an economical mushroom yield, the substrate at fill has to be short or dense enough to attain a high substrate dry weight.

Growing Systems (Phase II)

Once Phase I is complete, the substrate will be filled into a system for Phase II substrate preparation and to grow the mushrooms. Phase II takes place in one of three main types of mushroom-growing systems, depending on the type of production system used. The difference in the mushroom- growing systems is the container in which the crop is processed and grown. Each room has a different heating, ventilating, and airconditioning (HVAC) system designed for a specific stage in crop development. A single-zone system—or bed farm—consists of several large, stacked beds or shelves within a single room. The substrate is filled into these beds after Phase I, and the crop remains in the one room throughout the process. Bulk pasteurization or tunnels are systems where the substrate is filled into "tractortrailer"– type bins (called tunnels) with perforated floors and no covers on top of the compost. Phase II and, occasionally, the next phase of growing are carried out within these tunnels. The substrate may then be filled into a tray, shelf, or even plastic garbage bags for the remaining part of the process.

Phase II: Finishing the Compost

Phase II composting is the second step of compost substrate preparation. The first objective of Phase II is to pasteurize the composted substrate. The composted substrate is pasteurized to reduce or eliminate the bad microbes such as insects, other fungi, and bacteria. This is not a complete sterilization but a selective killing of pests that will compete for food or directly attack the mushroom. At the same time, this process minimizes the loss of good microbes.

The second goal of Phase II is to complete the composting process. Completing the composting process means eliminating all remaining simple soluble sugars and gaseous and soluble ammonia created during Phase I composting. Since ammonia is toxic to the mushroom mycelium, it must be converted to food the mushroom can use. The good microbes in Phase II convert toxic ammonia in solution and amine (other readily available nitrogen compounds) substances into protein—specific food for the mushroom. At the end of Phase II, volatile ammonia (concentration more than 0.05 percent) will inhibit mushroom spawn growth. Generally, ammonia concentrations above 0.10 percent can be easily detected by a person and are toxic to the spawn. Most of this conversion of ammonia and carbohydrates is accomplished by the growth of the microbes in the compost. These microbes are very efficient in using Phase I composting products, such as ammonia, as one of their main sources of food. The ammonia is incorporated as mostly protein into their bodies or cells. Eventually the mushroom uses these packets of nutrients as food. Phase II objectives are possibly the most difficult procedures in growing mushrooms. Because of a composting or other cultural problem, growers sometimes have to adjust Phase II programs. The Phase II process takes anywhere from 7 to 18 days, depending on how the air and compost temperatures are managed to control microbial activity.

During Phase II in standard bed or tray systems, compost temperatures are brought down through all temperature ranges to ensure that all the different species have a chance to convert their specific source of carbohydrates. The composted substrate throughout Phase II should appear to have moderate "firefang"—a term referring to the white-flecking microbial growth pattern of the thermophilic microorganism. Pasteurization (peak heat, boost) should be completed toward the start of Phase II. Effective pasteurization will eradicate harmful bacteria, nematodes, insects, and fungi. In general, air and composted substrate temperatures should be raised together to 140°F (60°C) for at least 2 hours. Growers make several compromises to this range, but it is a time-temperature relationship.

The good microbes grow best at temperatures from 115°F to 140°F; the more ammonia-utilizing microbes grow best in the temperature range of 120–128°F (47–49°C). The longer the microbes in the composted substrate remain in this optimum range with all the critical growth requirements available, the faster the ammonia will be converted. Understanding how these microbes grow and work in composted substrate should make the management of Phase II a little easier. The process of going through this temperature range will produce the most protein or the maximum amount of food for the mushroom. A good rule of thumb is not to drop the composted substrate temperature more than 5°F per 24 hours, which maintains the compost substrate in the desired range for about 4 or more days. Near the completion of Phase II, growers check for ammonia in the compost. The nose is usually the best tool. However, ammonia-testing kits and strips are available to supplement the nose test.

Spawn Maintenance

A desirable mycelial culture is pure—free of contaminants and of sectoring of other abnormalities. Contaminants include other fungi, bacteria, or insects growing on or infesting the culture media along with the desired mycelial culture. When a culture is first obtained, it should be transferred several times to fresh media to check for any form of contamination. Sectoring is any type of mycelial growth that differs in appearance, growth rate, colour, or in any other way from the typical appearance of a given strain. Sectoring is often observed as a more rapidly growing area near the leading edge of growth, exhibiting a different growth habit from the rest of the culture. Other abnormalities that might appear in a culture are fluffy, aerial mycelia, thick or rubbery textures, and colour changes such as browning or darkening of the mycelium. Sectors of other change in vegetative growth could affect the productivity of the culture. Therefore, recognizing and avoiding propagation of abnormal mycelia to agar and further spawn production is very important.

There is no in vitro test to determine a stock culture's validity. A series of cropping trials must be conducted on the mycelial stock culture to determine a culture line's value. Mushroom yield, size, colour, cap shape, and any other desired quality or growth factors are selected and then compared for each culture line. Many commercially prepared spawn strains are available to commercial and noncommercial growers. All commercially grown strains are pure culture of edible, fresh mushrooms; some may vary in texture and growing requirements.

Mushroom spawn is produced in several different strains or isolates. Hybrid White is a smooth-cap, highyield, excellent processing strain. Hybrid Off-White has a cap that is slightly scaly on first break and is a preferred fresh-market strain, and Brown (Portabella, Crimini) produces a chocolate-brown, mature mushroom that is fleshy and has a strong, mature flavour.

Spawn Production

The process of making spawn remains much the same as Penn State professor emeritus Dr. Sinden first developed in the 1930s. Grain is mixed with a little calcium carbonate, then cooked, sterilized, and cooled. Small pieces of pure-culture mycelium are placed in small batches on the grain. Once the small batch is fully colonized, it is used to inoculate several larger batches of grain. This multiplying of the inoculated grain continues until the commercial-size containers— usually plastic bags with breathable filter patches—are inoculated. During the colonization of each batch, the containers are shaken every few days to distribute actively growing mycelia around the bag or bottle. During the process, temperatures are maintained at 74–76°F (23–24°C). Uniformity of the air circulating around the bags is important to ensure that all containers are kept within the desired temperature range. Mycelium is sensitive and its fruiting mechanism can be easily damaged at high temperatures.

Spawning

On bed farms, spawn and supplement are broadcast over the surface of the substrate. Uniformity of this distribution is critical to achieve even spawn growth and temperatures. On tray or bulk farms, spawn is usually metered into the substrate during the mixing operation. Spawning is the cleanest operation performed on a mushroom farm. All equipment, baskets, tools, and so forth should be thoroughly cleaned and disinfected before spawning. The amount of spawn used depends on the length of the spawn-growing period and compost fill weights. The use of more spawn will result in a quicker colonization and more efficient use of substrate nutrients.

Improved colonization of substrate will help ensure that the mushroom mycelia will grow quicker than other fungal competitors.

During the spawn-growing period, heat is generated and supplemental cooling is required. Substrate temperatures should be maintained at 75–77°F and relative humidity should be high to minimize drying of the substrate surface. Under proper conditions, the spawn will grow as a delicate network of mycelia throughout the substrate. The mycelium grows in all directions from a spawn grain. Eventually mycelia from different spawn grains fuse together, making a spawned bed appear as a white root-like network throughout the compost. As the spawn grows, it generates heat. If the compost temperature increases to above 45 or 85°F, depending on the cultivar, the heat may kill or damage the mycelia, reducing crop yield and/or mushroom quality. The time needed for spawn to colonize the compost depends on the spawning rate and its distribution, the compost moisture and temperature, and the nature or quality of the compost. A complete spawn run usually requires 14 to 21 days. The spawn-growing period is considered complete when spawn has completely colonized the substrate and the metabolic heat surge is subsiding.

Substrate Supplementation

The compost has to provide the mushroom mycelium with a smorgasbord of food. Not only is ligninhumus complex and cellulose important, but protein, fat, and oils are also important. A good analogy is protein serves as the mushroom's "steak," carbohydrates its "potatoes," and lipids (fats and oils) its "butter." Like people, mushrooms should eat a balance of all these food types. The main source of "steak and butter" for the mushroom is from Phase II microbes. The dead cells of thermophilic fungi, bacteria, and actinomycetes "firefang" are the packages that deliver protein and fat to the mushroom.

The addition of delayed-release supplements further enhances the protein and lipid content of the compost for the mushroom. Many of these supplements consist of a high-protein oil material, such as soybean meal, cornmeal, or feather meal, that has been treated to delay the availability of the nutrient for the mushroom. If an untreated supplement is added to the compost at this time, it often becomes a "candy bar" to other microbes,weeds, or competitor molds. These molds grow more rapidly than the mushroom mycelium and can quickly colonize the compost, competing with the mushroom for nutrients.

The oils or lipids in these supplements are used by the mushroom to stimulate the fruiting mechanism and increase yield by having more mushrooms initiate and develop. Yields can be increased from 0.25 to 1.5 lbs/sq ft of growing space. In addition, mushroom size may also be improved in compost with higher spawning-moisture content. However, in substrate that is not selectively prepared, these nutrients become more available to com petitor molds. Often, if a farm is having composting problems, not supplementing until the problems are corrected is more economical.

Dairy Wastes and Treatment

The rapid growth in the size of dairy operations has resulted in new laws and regulations governing the handling and disposal of manure (Mitchell and Beddoes 2000). Requirements for nutrient management plans, manure solids disposal, and odour control (HouseBill 2001) make it necessary that new manure management approaches be considered. One of the more promising methods is anaerobic digestion.

Anaerobic digestion is a natural process that converts biomass to energy. Biomass is any organic material that comes from plants, animals or their wastes.

Anaerobic digestion has been used for over 100 years to stabilize municipal sewage and a wide variety of industrial wastes. Most municipal wastewater treatment plants use anaerobic digestion to convert waste solids to gas. The anaerobic process removes a vast majority of the odorous compounds (Lusk 1995),(Wilkie 2000),(Wilkie 2000). It also significantly reduces the pathogens present in the slurry (Lusk 1995).

Over the past 25 years, anaerobic digestion processes have been developed and applied to a wide array of industrial and agricultural wastes (Speece 1996), (Ghosh 1997). It is the preferred waste treatment process since it produces, rather than consumes, energy and can be carried out in relatively small, enclosed tanks. The products of anaerobic digestion have value and can be sold to offset treatment costs (Roos 1991).

Anaerobic digestion provides a variety of benefits. The environmental benefits include:

- Odors are significantly reduced or eliminated.
- Flies are substantially reduced.
- A relatively clean liquid for flushing and irrigation can be produced.
- Pathogens are substantially reduced in the liquid and solid products.
- Greenhouse gas emissions are reduced.
- And finally, nonpoint source pollution is substantially reduced. On the economic side, additional benefits are provided.
- The time devoted to moving, handling, and processing manure is minimized.
- Biogas is produced for heat or electrical power.
- Waste heat can be used to meet the heating and cooling requirements of the dairy.
- Concentrating nutrients to a relatively small volume for export from the site can reduce the land required for liquid waste application.
- The rich fertilizer can be produced for sale to the public, nurseries, or other crop producers.
- Income can be obtained from the processing of imported wastes (tipping fees), the sale of organic nutrients, greenhouse gas credits, and the sale of power.
- Power tax credits may be available for each kWh of power produced.
- Greenhouse tax credits may become available for each ton of carbon recycled.
- Finally the power generated is "distributed power" which minimizes the need to modify the power grid. The impact of new power on the power grid is minimized.

In order to achieve the benefits of anaerobic digestion, the treatment facility must be integrated into the dairy operation. Unfortunately, no single dairy can serve as a model for a manure treatment facility. The operation of the dairy will establish the digester loading and the energy generated from the system. The anaerobic facility must be designed to meet the individual characteristics of each dairy.

This manual provides an introduction to the anaerobic digestion of dairy manure. It is divided into three parts. The first describes the operation and waste management practices of Idaho dairies. The second introduces anaerobic digestion and the anaerobic digestion processes suitable for dairy waste. The third presents typical design applications for different types of dairies and establishes the cost and benefits of the facilities.

Dairy Operations

Dairy operations significantly affect the quantity and quality of manure that may be delivered to the anaerobic digestion system. In addition to the number of milk and dry cows, the housing, transport, manure separation, and bedding systems used by the dairy establishes the amount of material that must be handled and the amount of energy produced.

Housing System

Confined dairy animals may be housed in a variety of systems. Commonly used housing systems include free stalls, corrals with paved feed lanes, and open lot systems. Milk cows, dry cows and heifers may be housed in free stalls, corrals, and open-lots on the same dairy. The type of housing used determines the quantity of manure that can be economically collected.

Free Stall Barns

Free stalls are currently the most popular method for housing large dairy herds. Free stall housing provides a means for collecting essentially all of the manure.

Corrals

Corral systems with paved feed lanes are also commonly used. The manure deposited in the feed lanes can be scraped or flushed daily. From 40 to 55- percent of the excreted manure may be deposited and collected from the corral feed lane. The balance of the manure may be deposited in the milk barn (10 to 15 percent) or the open lot (30 to 50%).

Typically the manure deposited in the open lot is removed two to three times a year. It may have little net energy value after being stored in the open lot over prolonged periods of time.

For corral systems one must make a reasonable determination of the recoverable manure deposited in the feed lane, corral, and milk barn. Corral with Paved Feed Lane for Scrape or Vacuum Collection Corral systems also use a considerable amount of bedding material during the winter months. The straw bedding is generally removed in the spring and placed on the fields prior to spring planting.

Milk Barn

Dairy cows are milked two to three times a day. The cows are moved from their stalls to the milk parlor holding area. The milk parlor and holding area are normally flushed with fresh water. From 10 to 15 percent of the manure is deposited in the milk parlor. In addition to the manure that is flushed, the cows may be washed with a sprinkler system. Warm water that is produced by the refrigeration compressors, vacuum pumps, and milk cooling system may be used for drinking water, manure flushing or washing the cows. It has been estimated that 5 to 150 gallons of fresh water per milk cow is used in the milking centre. More common values are 10 to 30 gallons of fresh water per milk cow. The quantity and quality of water discharge from the milk parlor must be accurately measured. In many cases, the waste deposited in the milk barn is processed in a separate waste management system.

Open Lot

In open lot systems the manure is deposited on the ground and scraped into piles. The manure is removed infrequently (once or twice a year). A significant amount of manure degradation occurs resulting in greenhouse gas emissions. In many cases, the open lot degradation produces manure that has little or no net energy value.

Open Lot System

Transport System

The commonly used manure transport systems are flush, scrape, vacuum, and loader systems. In free stall barns the manure can be flushed, scraped, or vacuum collected.

Flush Systems

If a flush system is used the manure is substantially diluted. The quantity of water used in a flush system depends on the width, length, and slope of the flush isle. The feed isles are generally 14 feet wide while the back isles are generally 10 feet wide. The slope varies between one and two percent. A flush system will generally reduce the concentration of manure from 12 1/2 percent solids, "as excreted", to less than one percent solids in the flush water. Flush systems are however more economical and less labour-intensive than scrape or vacuum systems.

Scrape Systems

Scrape systems are simply systems that collect the manure by scraping it to a sump. Under normal weather conditions the scraped manure has approximately the same consistency as the "as excreted" manure. During the warm dry summer manure may be dewatered on the slab.

Front End Loader

Front-end loaders are used to stack and remove corral bedding and manure.

Vacuum Systems

Vacuum systems collect "as excreted" manure with a vacuum truck. Generally, the trucks collect approximately 4000 gallons per load. The manure can be hauled to a disposal site rather than to an intermediate sump. Vacuum collection is a slow and tedious process. The advantage is that the collected manure is undiluted and approximately equal to the "as excreted" concentration.

Bedding

The type of bedding used can significantly alter the characteristics of the manure being treated. Typically straw, wood chips, sand, or compost are used as bedding material. In some cases paper mixed with sawdust is used. Compost usually has some sand mixed with the organic constituents. If composting is carried out on dirt lots, a significant amount of sand and silt may be incorporated into the compost. Since anaerobic digestion will not degrade the wood chips, sand, or silt, it is necessary to remove those constituents prior to, or during anaerobic digestion process. The quantity of non-degradable, organic and inorganic material can significantly impact the performance of the anaerobic digester.

The quantity of bedding added to the manure is a function of the design and operation of the dairy. Generally only the "kick-out" from the stalls is added to the manure. The quantity that is "kicked-out" is a function of the design of the dairy housing system as well as the type of bedding used.

Manure Processing

Each dairy has its own manure processing system. Scraped or

flushed manure may be processed in a system separate from the milk barn waste, or the collected manure waste may be processed with the milk barn waste. In general, current manure processing consists of macerating the waste with a chopper pump, screening the waste to remove the organic fibers, followed by sedimentation to remove the sand, silt, and organic settable particles. Much of the degradable manure is removed during the separation processes. Up to 80% of the COD and 30% of the total Nitrogen and Phosphorous can be lost in the solids removed by the screen and sedimentation process. Detailed sampling and analysis is required to confirm losses.

Holding Tanks and Chopper Pumps

A wide variety of holding tanks and chopper pumps are used throughout the dairy industry. Typically, the tanks are relatively small but in some cases they are designed to hold several hours of flush water.

Anaerobic Digestion

Anaerobic digestion is the breakdown of organic material by a microbial population that lives in an oxygen free environment. Anaerobic means literally "without air". When organic matter is decomposed in an anaerobic environment the bacteria produce a mixture of methane and carbon dioxide gas. Anaerobic digestion treats waste by converting putrid organic materials to carbon dioxide and methane gas. This gas is referred to as biogas.

The biogas can be used to produce both electrical power and heat. The conversion of solids to biogas results in a much smaller quantity of solids that must be disposed. During the anaerobic treatment process, organic nitrogen compounds are converted to ammonia, sulfur compounds are converted to hydrogen sulfide, phosphorus to orthophosphates, and calcium, magnesium, and sodium are converted to a variety of salts.

Through proper operation, the inorganic constituents can be converted to a variety of beneficial products. The end products of anaerobic digestion are natural gas (methane) for energy production, heat produced from energy production, a nutrient rich organic slurry, and other marketable inorganic products.

The effluent containing particulate and soluble organic and inorganic materials can be separated into its particulate and soluble constituents. The particulate solids can be sold or exported from the dairy while the nutrient rich liquids are applied to the land.

Bacterial Consortia

Anaerobic digestion is carried out by a group, or consortia of bacteria, working together to convert organic matter to gas and inorganic constituents. The first step of anaerobic digestion is the breakdown of particulate matter to soluble organic constituents that can be processed through the bacterial cell wall. Hydrolysis, or the liquification of insoluble materials is the rate-limiting step in anaerobic digestion of waste slurries.

This step is carried out by a variety of bacteria through the release of extra-cellular enzymes that reside in close proximity to the bacteria. The soluble organic materials that are produced through hydrolysis consist of sugars, fatty acids, and amino acids. Those soluble constituents are converted to carbon dioxide and a variety of short chain organic acids by acid forming bacteria. Other groups of bacteria reduce the hydrogen toxicity by scavenging hydrogen to produce ammonia, hydrogen sulfide, and methane. A group of methanogens converts acetic acid to methane gas. A wide variety of physical, chemical, and biological reactions take place. The bacterial consortia catalyze these reactions. Consequently, the most important factor in converting waste to gas is the bacterial consortia. The bacterial consortia are essentially the "bio-enzymes" that accomplish the desired treatment. A poorly developed or stressed bacterial consortium will not provide the desired conversion of waste to gas and other beneficial products.

Factors Controlling the Conversion of Waste to Gas

The rate and efficiency of the anaerobic digestion process is controlled by:

- The type of waste being digested,
- Its concentration,
- Its temperature,
- The presence of toxic materials,
- The pH and alkalinity,
- The hydraulic retention time,
- The solids retention time,
- The ratio of food to microorganisms,
- The rate of digester loading,
- And the rate at which toxic end products of digestion are removed.

Waste Characteristics

All waste constituents are not equally degraded or converted to gas through anaerobic digestion. Anaerobic bacteria do not degrade lignin and some other hydrocarbons. The digestion of waste containing high nitrogen and sulfur concentrations can produce toxic concentrations of ammonia and hydrogen sulfide. Wastes that are not particularly water-soluble will breakdown slowly. Dairy wastes have been reported to degrade slower than swine or poultry manure. As pointed out earlier, lignin will not degrade during anaerobic digestion. Since a substantial portion of the volatile solids in dairy waste is lignin, the percentage of cow manure volatile solids that can be converted to gas is lower when compared to other manure and wastes. The manure characteristics also establish the percentage of carbon dioxide and methane in the biogas produced. Dairy waste biogas will typically be composed of 55 to 65% methane and 35 to 45% carbon dioxide. Trace quantities of hydrogen sulfide and nitrogen will also be present.

Dilution of Waste

The waste characteristics can be altered by simple dilution. Water will reduce the concentration of certain constituents such as nitrogen and sulfur that produce products (ammonia and hydrogen sulfide) that are inhibitory to the anaerobic digestion process. High solids digestion creates high concentrations of end products that inhibit anaerobic decomposition. Therefore, some dilution can have positive effects. The literature indicates that greater reduction efficiencies occur at concentrations of approximately 6 to 7 percent total solids. Dairy waste "as excreted" is approximately 12 percent total solids and 10.5 percent volatile solids. Most treatment systems operate at a lower solids concentration than the "as excreted" values.

Component	% Dry of Matter
Volatile solids	83.0
Ether Extract	2.6
Cellulose	31.0
Hemicellulose	12.0
Lignin	12.2
Starch	12.5
Crude Protein	12.5
Ammonia	0.5
Acids	0.1

Dilution also causes stratification within the digester. Undigested straw forms a thick mat on top of the digester while sand accumulates at the bottom. The optimum waste concentration is based on temperature and the quantity of straw and other constituents that are likely to separate within the anaerobic digester. It is desirable to keep the separation or stratification in the digester to a minimum. Intense mixing involving the consumption of power may reduce the stratification of dilute waste.

The use of flush systems to remove the manure from the dairy barns has major economic advantages to the dairy. Flush systems normally use 100 to 200 gallons per cow, per day of dilution water. The flush volumes required are based on the lane or gutter length, width and slope (Fulhage and Martin 1994). The flush water usually contains very low concentrations of total and volatile solids. At 100 gallons per cow of flush water, the waste has only 12.5 percent of the "as excreted" concentration. At 200 gallons per cow per day of flush water the waste contains only 6.25 percent of the "as excreted" concentration. The milk parlor also produces a substantial amount of dilute waste. Approximately 15% of the animal manure is deposited in the milk parlor.

Foreign Materials

Addition of foreign materials such as animal bedding, sand and silt can have a significant impact on the anaerobic digestion process. For example, the poor performance of the Monroe, WA dairy digester was attributed to the use of cedar wood chip bedding (Ecotope 1979). The quantity and quality of the bedding material added to the manure will have a significant impact on the anaerobic digestion of dairy waste. Sand and silt must be removed before anaerobic digestion. If it is not removed before digestion it must be suspended during the digestion process.

Toxic Materials

Toxic materials such as fungicides and antibacterial agents can have an adverse effect on anaerobic digestion. The anaerobic process can handle small quantities of toxic materials without difficulty. Storage containers for fungicides and antibacterial agents should be placed at locations that will not discharge to the anaerobic digester.

Nutrients

Bacteria require a sufficient concentration of nutrients to achieve optimum growth. The carbon to nitrogen ratio in the waste should be less than 43. The carbon to phosphorus ratio should be less than 187. Hills and Roberts showed that a non-lignin C/N ratio of 20 to 25 is optimum for digester performance. Typically, "as excreted manure has a C/N ratio of 10.

Temperature

The anaerobic bacterial consortia function under three temperature ranges.

Psychrophilic temperatures of less than 68 degrees Fahrenheit produce the least amount of bacterial action. Mesophilic digestion occurs between 68 degrees and 105 degrees Fahrenheit. Thermophilic digestion occurs between 110 degrees Fahrenheit and 160 degrees Fahrenheit. The optimum mesophilic temperature is between 95 and 98 degrees Fahrenheit. The optimum thermophilic temperature is between 140 and 145 degrees Fahrenheit. The rate of bacterial growth and waste degradation is faster under thermophilic conditions. On the other hand, thermophilic digestion produces an odorous effluent when compared to mesophilic digestion. Thermophilic digestion substantially increases the heat energy required for the process. In most cases, sufficient heat is not available to operate in the thermophilic range. This is especially true if flush systems are used or the milk parlor waste is mixed with the scraped manure. Large quantities of dilution flush water must be heated to the digester's operating temperature.

During cold weather, control of the flush volume is critical in maintaining adequate digester temperatures. Seasonal and diurnal temperature fluctuations significantly affect anaerobic digestion and the quantities of gas produced. Bacterial storage and operational controls must be incorporated in the process design to maintain process stability under a variety of temperature conditions.

Temperature is a universal process variable. It influences the rate of bacterial action as well as the quantity of moisture in the biogas. The biogas moisture content increases exponentially with temperature. Temperature also influences the quantity of gas and volatile organic substances dissolved in solution as well as the concentration of ammonia and hydrogen sulfide gas.

pH

Methane producing bacteria require a neutral to slightly alkaline environment (pH 6.8 to 8.5) in order to produce methane. Acid forming bacteria grow much faster than methane forming bacteria. If acid-producing bacteria grow too fast, they may produce more acid than the methane forming bacteria can consume. Excess acid builds up in the system. The pH drops, and the system may become unbalanced, inhibiting the activity of methane forming bacteria. Methane production may stop entirely. Maintenance of a large active quantity of methane producing bacteria prevents pH instability. Retained biomass systems are inherently more stable than bacterial growth based systems such as completely mixed and plug flow digesters.

Hydraulic Retention Time (HRT)

Most anaerobic systems are designed to retain the waste for a fixed number of days. The number of days the materials stays in the tank is called the Hydraulic Retention Time or HRT. The Hydraulic Retention Time equals the volume of the tank divided by the daily flow (HRT=V/Q). The hydraulic retention time is important since it establishes the quantity of time available for bacterial growth and subsequent conversion of the organic material to gas. A direct relationship exists between the hydraulic retention time and the volatile solids converted to gas.

Solids Retention Time (SRT)

The Solids Retention Time (SRT) is the most important factor controlling the conversion of solids to gas. It is also the most important factor in maintaining digester stability. Although the calculation of the solids retention time is often improperly stated, it is the quantity of solids maintained in the digester divided by the quantity of solids wasted each day.

Where V is the digester volume; Cd is the solids concentration in the digester; Qw is the volume wasted each day and Cw is the solids concentration of the waste. In a conventional completely mixed, or plug flow digester, the HRT equals the SRT. However, in a variety of retained biomass reactors the SRT exceeds the HRT. As a result, the retained biomass digesters can be much smaller while achieving the same solids conversion to gas.

The volatile solids conversion to gas is a function of SRT (Solids Retention Time) rather than HRT. At a low SRT sufficient time is not available for the bacteria to grow and replace the bacteria lost in the effluent. If the rate of bacterial loss exceeds the rate of bacteria growth, "wash-out" occurs. The SRT at which "wash-out" begins to occur is the "critical SRT".

Jewel established that a maximum of 65 percent of dairy manure's volatile solids could be converted to gas with long solids retention times. Burke established that 65 to 67 percent of dairy manure COD could be converted to gas. Long retention times are required for the conversion of cellulose to gas.

The goal of process engineers over the past twenty years has been to develop anaerobic processes that retain biomass in a variety of forms such that the SRT can be increased while the HRT is decreased. The goal has been to retain, rather than waste the biocatalyst (bacterial consortia) responsible for the anaerobic process. As a result of this effort, gas yields have increased and digester volumes decreased. A measure of the success of biomass retention is the SRT/HRT ratio. In conventional digesters, the ratio is 1.0. Effective retention systems will have SRT/HRT ratios exceeding 3.0. At an SRT/HRT ratio of 3.0 the digester will be 1/3rd the size of a conventional digester.

Digester Loading (kg / m3 / d)

Neither the hydraulic retention time (HRT), nor the solids retention time (SRT) tells the full story of the impact that the influent waste concentration has on the anaerobic digester. One waste may be dilute and the other concentrated. The concentrated waste will produce more gas per gallon and affect the digester to a much greater extent than the diluted waste. A more appropriate measure of the waste on the digester's size and performance is the loading. The loading can be reported in pounds of waste (influent concentration x influent flow) per cubic foot of digester volume. The more common units are kilograms of influent waste per cubic meter of digester volume per day (kg / m3 / d). One (kg / m3 / d) is equal to 0.0624 (lb / ft3 / d). The digester loading can be calculated if the HRT and influent waste concentration are known.

Food to Microorganism Ratio

The food to microorganism ratio is *the key factor* controlling anaerobic digestion. At a given temperature, the bacterial consortia can only consume a limited amount of food each day. In order to consume the required number of pounds of waste one must supply the proper number of pounds of bacteria. The ratio of the pounds of waste supplied to the pounds of bacteria available to consume the waste is the food to microorganism ratio (F/M). This ratio is the controlling factor in all biological treatment processes. A lower the F/M ratio will result in a greater percentage of the waste being converted to gas. Unfortunately, the bacterial mass is difficult to measure since it is difficult to differentiate the bacterial mass from the influent waste. The task would be easier if *all* of the influent waste were converted to biomass or gas. In that case, the F/M ratio would simply be the digester loading divided by the concentration of volatile solids (biomass) in the digester (L / Cd). For any given loading, the efficiency can be improved by lowering the F/M ratio by *increasing the concentration of biomass in the digester*. Also for any given biomass concentration within the digester, the efficiency can be improved by decreasing the loading. Unfortunately, a portion of the influent waste is not processed or converted to biomass or gas by the bacteria. In that case the F/M ratio is equal to the VS loading divided by the digester VS measured (VSD) minus the unprocessed Volatile Solids (VSUP). The unprocessed volatile solids may include refractory or non-degradable biological products produced by the bacteria.

End Product Removal

The end products of anaerobic digestion can adversely affect the digestion process. Such products of anaerobic digestion include organic acids, ammonia nitrogen, and hydrogen sulfide.

For any given volatile solids conversion to gas, the higher the influent waste concentration, the greater the end product concentration. End product inhibition can be reduced by lowering the influent waste concentration or by separately removing the soluble end products from the digester through elutriation. Elutriation is the process of washing the solids (bacteria) with clean water to remove the products of digestion. The contact process provides an efficient means of removing the end products of digestion. End product removal can be enhanced by elutriation, which is easily incorporated into the contact process (Burke 1997).

Digester Types

A vast array of anaerobic digesters have been developed and placed in operation over the past fifty years. A variety of schemes could be used to classify the digestion processes. For dairy waste, the most important classification is whether or not it can be used to convert dairy waste solids to gas while meeting the goals of anaerobic digestion. The goals of dairy waste anaerobic digestion are as follows:

1. Reduce the mass of solids
2. Reduce the odors associated with the waste products
3. Produce clean effluent for recycle and irrigation
4. Concentrate the nutrients in a solid product for storage or export

5. Generate energy

6. Reduce pathogens associated with the waste.

In addition, the digester must be able to handle or process the dairy waste stream. Dairy waste is a semi-solid slurry. Much of the energy value is in the solids. Consequently, the process must be able to convert solids to gas without clogging the anaerobic reactor. The process must also be able to handle bedding material, sand and other foreign materials associated with typical dairy waste. In addition, if the dairy manure is a dilute waste, the process must be capable of mitigating stratification and solids separation within the reactor.

Processes that are not Appropriate for Digesting Dairy Manure

A variety of high rate anaerobic processes, which retain bacteria have been developed to treat soluble organic industrial wastes. These "high rate" digesters have reduced hydraulic detention times from 20 days to a few hours. They include anaerobic filters, both upflow and downflow, and a variety of biofilm processes such as fixed film packed bed reactors. Bacteria are retained in these reactors as films on carriers such as plastic beads, or sand, or on support media of all configurations. The waste washes past the retained bacteria. The bacteria convert the soluble constituents to gas but have little opportunity to hydrolyze and degrade the particulate solids, unless the solids become attached to the biomass.

These reactors are *not* suitable for digesting dairy waste since they are not effective in converting particulate solids to gas and tend to clog while digesting dairy manure slurries. These high rate reactors can treat the soluble component of dairy waste. But only a fraction of the available energy will be recovered.

A widely used industrial waste anaerobic digester is the UASB or "Upflow Anaerobic Sludge Blanket", reactor. The process stores the anaerobic consortia as pellets, approximately the size of a pea. The upflow anaerobic sludge blanket reactor (UASB) is widely used in industrial treatment processes throughout the world. It is an extremely effective process for converting soluble organic materials, such as sugar to methane gas. It has not been used for processing dairy waste since it is ineffective in converting solids to gas. It is primarily used to convert non-particulate or soluble waste to gas.

UASB Reactor

The anaerobic baffled reactor is a horizontal version of the upflow anaerobic sludge blanket reactor. Both store large quantities of anaerobic bacteria as pellets approximately the size of a pea. Unfortunately, these very successful anaerobic reactors are not effective in digesting particulate waste. Particulate solids tend to settle in the horizontal baffled reactor (HBR) while organic fibers will form a mat on the surface. There are no known instances of the HBR being used for the treatment of dairy waste. Unless the dairy waste was thoroughly screened and all particulate matter removed the HBR would tend to become clogged. The removal of solids by screening and gravity sedimentation will eliminate up to 80% of the energy generating potential from dairy waste.

Processes that can be used for Digesting Dairy Manure

The processes that have been used for digesting dairy waste can be subdivided into high rate and low rate processes. Low rate processes consist of covered anaerobic lagoons, plug flow digesters, and mesophilic completely mixed digesters. High rate reactors include the thermophilic completely mixed digesters, anaerobic contact digesters, and hybrid contact/fixed film reactors.

Anaerobic Lagoons (Very Low Rate)

Anaerobic lagoons are covered ponds. Manure enters at one end and the effluent is removed at the other. The lagoons operate at psychrophilic, or ground temperatures. Consequently, the reaction rate is affected by seasonal variations in temperature.

Best Management Practices to Handle Dairy Wastes

Dairy farmers recognize the importance of water quality; they accept the responsibility of managing their dairy wastes to protect, preserve, and even improve the quality of both surface and ground waters. No single dairy waste management practice can meet the needs of every dairy nor is a single waste management practice appropriate or practical on all dairies in Alabama.

Wise dairy operators realize that the best answer to their situation will be a combination of waste management practices. Such dairy operators can seek input from several sources. Extension dairy production and waste management specialists and waste management and water quality specialists from NRCS and other agricultural agencies can help operators establish the best waste management system for their dairy.

This publication briefly describes various animal waste management practices. When properly applied to specific dairy situations, these activities become "best management practices" that help dairy operators:

- Prevent direct discharge of manure or wastewater into surface waters or onto adjacent neighbours' property.
- Prevent any nuisance conditions that interfere with normal use and enjoyment of neighbours' property.
- Enhance the operational efficiency of the dairy unit.
- Collect and use dairy manure and wastewater for beneficial purposes such as fertilizer, compost, or bedding.

Dairy operators in Alabama can use one or more of the following dairy waste management techniques to reach the goals stated above.

Holding Pond for Milking Centre Wastes

Dairy wastes associated with the milking centre are liquid and are easily collected in a holding pond close to the milking centre. This holding pond can also collect liquid wastes and runoff from holding areas. The pond should be sized to hold this liquid waste plus any rainfall that occurs between pond pump-out intervals.

Runoff Management

Noncattle open areas around the dairy milking centre or dairy barn facility should be sloped to direct freshwater runoff away from cattle areas and manure collection structures unless additional water is needed for irrigation.

Unless sized for this additional runoff, ponds and lagoons will fill and require pumping out quicker than expected. For wastewater irrigation, all runoff can be directed into either the milking centre holding pond, waste storage pond, or lagoon system. This allows for land application with other collected dairy waste. Roof guttering is one way of collecting roof runoff and controlling where it goes, either away from or into the waste stream as desired.

Scrape and Haul

This system uses a scraping operation to remove solid manure from holding areas and even some freestall barns to a special manure holding area to await hauling for land application. This special manure holding area should have an impervious base such as compacted clay, but concrete is preferable. This impervious base should be sloped to drain liquid manure away into a waste storage pond that also handles milking centre liquid wastes. Storage areas may also be covered to reduce rainfall runoff into the system. This method of dairy waste management is very basic and is generally most applicable for dairy herds with less than 100 cows. In some instances, the manure hauling and land application interval could be daily, but more than likely it will be weekly or monthly.

Waste Storage Pond

A dairy waste storage pond is a specially constructed pond used to collect and store manure, flush water, and polluted runoff from a dairy facility. Storage is a relatively short period of 90 to 180 days. Waste storage pond contents must be removed at the end of this storage period with land application the most common end use. The waste storage pond must be constructed with an impermeable liner to prevent waste leakage to ground water. Because waste storage ponds produce an odour, they should be located downwind of neighbours, highways, and any public use area. Storage pond waste retains most of its fertilizer nutrients. As a result, at a land application rate of 200 pounds of nitrogen per acre, more than two and one-half times as much cropland will be needed for land application of waste from a storage pond system than from a two-stage lagoon system for the same number of dairy cows.

Dairy Waste Lagoon

Dairy waste lagoons are earthen structures designed for biological treatment and long-term storage of dairy waste. Lagoons are specially constructed to prevent leakage of dairy waste to ground water. The lagoon system allows manure to be handled with water-flushing systems, sewer lines, pumps, and irrigation equipment. The natural biological action on the waste results in less odour during land application. Nitrogen content of the waste is reduced in lagoons by as much as 80 percent.

This reduction minimizes land area needed for land application and enhances long-term storage.

Waste Storage Pond Agitation and Lagoon Renovation

Agitation prior to and during both waste storage pond pumpout and lagoon renovation is necessary to suspend waste solids that have settled to the bottom or floated to the surface.

This agitation will be by either a portable propeller-type pump or a chopper-type pump requiring at least a 100-horsepower tractor. Agitation should thoroughly mix the solid and liquid contents to allow removal of both during the pump-out process.

Solid/Liquid Separation for Dairy Waste

Twelve to 14 percent of excreted dairy manure is solids. This, along with any solid bedding material used in freestall housing, will reduce storage pond or lagoon capacity. Even after biological treatment, the lagoon will be quickly filled.

These accumulated solids also greatly reduce lagoon efficiency. Solid-liquid separation using either settling basins or mechanical separators (stationary screens or elevators) can reduce up to half of the solids in the waste stream and should be considered even when no bedding is used. These solids can then be used for fertilizer, bedding, feed, or composting. If used, separators can increase the capacity of storage ponds and waste lagoons.

Land Application

Land application is the best end use of dairy manure collected regardless of the handling method. Solids are applied with manure spreaders, and liquid contents of storage ponds and dairy lagoons are land applied through irrigation. Land application should match fertilizer requirements of the target crop with plant-available nutrients in the waste. Matching crop fertilizer requirements minimizes harmful effects to water quality that can occur through runoff into surface water or deep percolation into ground water. It also reduces the risk of water pollution from manure application to the same level as manufactured fertilizer application. Timing of applications and soil incorporation should also be considered.

Waste Testing

Proper land application of waste requires nutrient content of the waste to be analyzed or estimated. While laboratory analysis of lagoon wastewater and agitated lagoon slurry is available, a 3- to 10-day time lag between sampling and receiving analysis results is typical. Field tests for plant-available nitrogen are available to allow on-site wastewater and dairy slurry testing during the land application process. This allows calibration of the land application process to provide the proper rate of nutrient application for a crop.

Soil Testing

Prior to land application of dairy wastewater or slurry waste, the target field intended for waste application should be soil tested. Soil testing tells dairy operators how much fertilizer is required for the crop and allows them to match the dairy waste/wastewater application to meet these crop fertilizer requirements.

Wastewater Irrigation

Although tractor-pulled tanks with injection of wastes are available, they are not practical for dairies in Alabama.

Using regular agricultural irrigation pumps and either travelling guns or solid set or centre pivot irrigation equipment to land apply dairy wastewater (liquid with less than 2 percent solids content) is the best way to handle large volumes of liquid wastes generated by a freestall flush system. Irrigation equipment should be calibrated to match crop fertilizer requirements.

Stream Exclusion

In those pasture areas where dairy animals have access to flowing streams, steps should be taken to minimize cattle loitering in the stream. Heavy or high-use access could cause stream bank erosion or destruction of stream side vegetation and water quality reduction due to manure deposits. Stream shade removal, fences, and other barriers may be used to protect the quality of streams flowing through these pastures.

Rotational Grazing

Dividing pastureland, either rainfed or irrigated, into small areas for controlled daily grazing allows opportunity for greater per feed unit milk production through intensive pasture management for grazing. Direct land application of dairy waste and wastewater promotes grass production and easier manure handling.

Chapter 6

The More Common Dairy Processes

The dairy industry is divided into two main production areas:

- the primary production of milk on farms—the keeping of cows (and other animals such as goats, sheep etc.) for the production of milk for human consumption;

- the processing of milk—with the objective of extending its saleable life. This objective is typically achieved by (a) heat treatment to ensure that milk is safe for human consumption and has an extended keeping quality, and (b) preparing a variety of dairy products in a semi-dehydrated or dehydrated form (butter, hard cheese and milk powders), which can be stored.

The focus of this document is on the processing of milk and the production of milk-derived products—butter, cheese and milk powder—at dairy processing plants. The upstream process of primary milk production on dairy farms is not covered, since this activity is more related to the agricultural sector. Similarly, downstream processes of distribution and retail are not covered.

Dairy processing occurs world-wide; however the structure of the industry varies from country to country. In less developed countries, milk is generally sold directly to the public, but in major milk producing countries most milk is sold on a wholesale basis. In Ireland and Australia, for example, many of the large-scale processors are owned by the farmers as co-operatives, while in the United States individual contracts are agreed between farmers and processors.

Dairy processing industries in the major dairy producing countries have undergone rationalisation, with a trend towards fewer but larger plants operated by fewer people. As a result, in the United States, Europe, Australia and New Zealand most dairy processing plants are quite large. Plants producing market milk and products with short shelf life, such as yogurts, creams and soft cheeses, tend to be located on the fringe of urban centres close to consumer markets. Plants manufacturing items with longer shelf life, such as butter, milk powders, cheese and whey powders, tend to be located in rural areas closer to the milk supply.

The general tendency world-wide, is towards large processing plants specialising in a limited range of products. There are exceptions, however. In eastern Europe for example, due to the former supply-driven concept of the market, it is still very common for 'city' processing plants to be large multi-product plants producing a wide range of products.

The general trend towards large processing plants has provided companies with the opportunity to acquire bigger, more automated and more efficient equipment. This technological development has, however, tended to increase environmental loadings in some areas due to the requirement for long-distance distribution.

Basic dairy processes have changed little in the past decade. Specialised processes such as ultrafiltration (UF), and modern drying processes, have increased the opportunity for the recovery of milk solids that were formerly discharged. In addition, all processes have become much more energy efficient and the use of electronic control systems has allowed improved processing effectiveness and cost savings.

Process Overview

Milk Production

The processes taking place at a typical milk plant include:
- receipt and filtration/clarification of the raw milk;
- separation of all or part of the milk fat (for standardisation of market milk, production of cream and butter and other fat-based products, and production of milk powders);
- pasteurisation;
- homogenisation (if required);
- deodorisation (if required);
- further product-specific processing;
- packaging and storage, including cold storage for perishable products;
- distribution of final products.

In such plants, yogurts and other cultured products may also be produced from whole milk and skimmed milk.

Butter Production

The butter-making process, whether by batch or continuous methods, consists of the following steps:

- preparation of the cream;
- destabilisation and breakdown of the fat and water emulsion;
- aggregation and concentration of the fat particles;
- formation of a stable emulsion;
- packaging and storage;
- distribution.

Is a flow diagram outlining the basic processing system for a butter-making plant. The initial steps, (filtration/clarification, separation and pasteurisation of the milk) are the same as described in the previous section. Milk destined for butter making must not be homogenised, because the cream must remain in a separate phase. After separation, cream to be used for butter making is heat treated and cooled under conditions that facilitate good whipping and churning. It may then be ripened with a culture that increases the content of diacetyl, the compound responsible for the flavour of butter.

Alternatively, culture inoculation may take place during churning. Butter which is flavour enhanced using this process is termed lactic, ripened or cultured butter. This process is very common in continental European countries. Although the product is claimed to have a superior flavour, the storage life is limited. Butter made without the addition of a culture is called sweet cream butter. Most butter made in the English-speaking world is of this nature. Both cultured and sweet cream butter can be produced with or without the addition of salt. The presence of salt affects both the flavour and the keeping quality.

Butter is usually packaged in bulk quantities (25 kg) for long-term storage and then re-packed into marketable portions (usually 250 g or 500 g, and single-serve packs of 10–15 g). Butter may also be packed in internally lacquered cans, for special markets such as the tropics and the Middle East.

Cheese Production

Virtually all cheese is made by coagulating milk protein (casein) in a manner that traps milk solids and milk fat into a curd matrix.

This curd matrix is then consolidated to express the liquid fraction, cheese whey. Cheese whey contains those milk solids which are not held in the curd mass, in particular most of the milk sugar (lactose) and a number of soluble proteins.

Milk Powder Production

Milk used for making milk powder, whether it be whole or skim milk, is not pasteurised before use. The milk is preheated in tubular heat exchangers before being dried. The preheating temperature depends on the season (which affects the stability of the protein in the milk) and on the characteristics desired for the final powder product. The preheated milk is fed to an evaporator to increase the concentration of total solids. The solids concentration that can be reached depends on the efficiency of the equipment and the amount of heat that can be applied without unduly degrading the milk protein.

The milk concentrate is then pumped to the atomiser of a drying chamber. In the drying chamber the milk is dispersed as a fine fog-like mist into a rapidly moving hot air stream, which causes the individual mist droplets to instantly evaporate. Milk powder falls to the bottom of the chamber, from where it is removed. Finer milk powder particles are carried out of the chamber along with the hot air stream and collected in cyclone separators.

Milk powders are normally packed and distributed in bulk containers or in 25 kg paper packaging systems. Products sold to the consumer market are normally packaged in cans under nitrogen. This packaging system improves the keeping quality, especially for products with high fat content.

Environmental Impacts

This section briefly describes some of the environmental impacts associated with the primary production of milk and the subsequent processing of dairy products. While it is recognised that the primary production of milk has some significant environmental impacts, this document is predominantly concerned with the processing of dairy products.

Impacts of Primary Production

The main environmental issues associated with dairy farming are:

- the generation of solid manure and manure slurries, which may pollute surface water and groundwater;

- the use of chemical fertilisers and pesticides in the production of pastures and fodder crops, which may pollute surface water and groundwater;
- the contamination of milk with pesticides, antibiotics and other chemical residues.

In most cases, solid manure is applied to pastures and cultivated land. The extent of application, however, may be restricted in some regions. Dairy effluent and slurries are generally held in some form of lagoon to allow sedimentation and biological degradation before they are irrigated onto land. Sludge generated from biological treatment of the dairy effluent can also be applied to pastures, as long as it is within the allowable concentrations for specified pollutants, as prescribed by regulations. Sludge can also be used in the production of methane-rich biogas, which can then be used to supplement energy supplies. Manure waste represents a valuable source of nutrients. However improper storage and land application of manure and slurries can result in serious pollution of surface waters and groundwater, potentially contaminating drinking water supplies. The extensive use of chemical fertilisers containing high levels of nitrogen has resulted in pollution of the groundwater and surface waters in many countries.

Nitrite in drinking water is known to be carcinogenic, and nitrite levels in drinking water that exceed 25–50 mg/L have been linked to cyanosis in newborn infants ('blue babies').

Compounds containing nitrogen and phosphorus, if discharged to surface water, can lead to excessive algal growth (eutrophication). This results in depleted dissolved oxygen levels in the water, thereby causing the death of fish and other aquatic species. In sensitive areas, therefore, the rate and manner of application of chemical fertilisers are critical.

The use of pesticides has been recognised as an environmental concern for many agricultural activities. Toxic pesticides, some of which biodegrade very slowly, can accumulate in body tissues and are harmful to ecosystems and to human health. Pesticides can end up in agricultural products, groundwater and surface waters, and in extreme cases can enter the human food chain through milk.

For the past few decades, the contamination of milk with antibiotics has been an issue of concern. This is due to the overuse of antibiotics for treatment of cattle diseases, particularly mastitis. It has been brought under control in most countries with developed dairy industries, through strict limitations on the use of antibiotics, regular testing of milk for antibiotic residues, rigorous enforcement of regulations, and education.

In some countries, considerable attention has also been paid to the screening of milk supplies for traces of radioactivity, and most countries now apply acceptance limits for raw and imported milk products. Even the slightest levels of contamination in milk can be serious, because pollutants are concentrated in the processing process.

Impacts of Dairy Processing

As for many other food processing operations, the main environmental impacts associated with all dairy processing activities are the high consumption of water, the discharge of effluent with high organic loads and the consumption of energy. Noise, odour and solid wastes may also be concerns for some plants.

Dairy processing characteristically requires very large quantities of fresh water. Water is used primarily for cleaning process equipment and work areas to maintain hygiene standards.

The dominant environmental problem caused by dairy processing is the discharge of large quantities of liquid effluent. Dairy processing effluents generally exhibit the following properties:

- high organic load due to the presence of milk components;
- fluctuations in pH due to the presence of caustic and acidic cleaning agents and other chemicals;
- high levels of nitrogen and phosphorus;
- fluctuations in temperature.

If whey from the cheese-making process is not used as a by-product and discharged along with other wastewaters, the organic load of the resulting effluent is further increased, exacerbating the environmental problems that can result.

In order to understand the environmental impact of dairy processing effluent, it is useful to briefly consider the nature of milk. Milk is a complex biological fluid that consists of water, milk fat, a number of proteins (both in suspension and in solution), milk sugar (lactose) and mineral salts.

Dairy products contain all or some of the milk constituents and, depending on the nature and type of product and the method of manufacturing, may also contain sugar, salts (e.g. sodium chloride), flavours, emulsifiers and stabilisers. For plants located near urban areas, effluent is often discharged to municipal sewage treatment systems. For some municipalities, the effluent from local dairy processing plants can represent a significant load on sewage treatment plants. In extreme cases, the organic load of waste milk solids entering a sewage system may well exceed that of the township's domestic waste, overloading the system.

In rural areas, dairy processing effluent may also be irrigated to land. If not managed correctly, dissolved salts contained in the effluent can adversely affect soil structure and cause salinity. Contaminants in the effluent can also leach into underlying groundwater and affect its quality.

In some locations, effluent may be discharged directly into water bodies. However this is generally discouraged as it can have a very negative impact on water quality due to the high levels of organic matter and resultant depletion of oxygen levels.

Electricity is used for the operation of machinery, refrigeration, ventilation, lighting and the production of compressed air. Like water consumption, the use of energy for cooling and refrigeration is important for ensuring good keeping quality of dairy products and storage temperatures are often specified by regulation. Thermal energy, in the form of steam, is used for heating and cleaning. As well as depleting fossil fuel resources, the consumption of energy causes air pollution and greenhouse gas emissions, which have been linked to global warming.

Dairy products such as milk, cream and yogurt are typically packed in plastic-lined paperboard cartons, plastic bottles and cups, plastic bags or reusable glass bottles. Other products, such as butter and cheese, are wrapped in foil, plastic film or small plastic containers. Milk powders are commonly packaged in multi-layer kraft paper sacs or tinned steel cans, and some other products, such as condensed milks, are commonly packed in cans.

Breakages and packaging mistakes cannot be totally avoided. Improperly packaged dairy product can often be returned for reprocessing; however the packaging material is generally discarded.

Emissions to air from dairy processing plants are caused by the high levels of energy consumption necessary for production. Steam, which is used for heat treatment processes (pasteurisation, sterilisation, drying etc.) is generally produced in on-site boilers, and electricity used for cooling and equipment operation is purchased from the grid. Air pollutants, including oxides of nitrogen and sulphur and suspended particulate matter, are formed from the combustion of fossil fuels, which are used to produce both these energy sources. In addition, discharges of milk powder from the exhausts of spray drying equipment can be deposited on surrounding surfaces. When wet these deposits become acidic and can, in extreme cases, cause corrosion.

For operations that use refrigeration systems based on chlorofluorocarbons (CFCs), the fugitive loss of these gases to the atmosphere is an important environmental consideration, since CFCs are recognised to be a cause of ozone depletion in the atmosphere. For such operations, the replacement of CFC-based systems with non-or reduced-CFC systems is thus an important issue.

Some processes, such as the production of dried casein, require the use of hammer mills to grind the product. The constant noise generated by this equipment has been known to be a nuisance in surrounding residential areas. The use of steam injection for heat treatment of milk and for the creation of reduced pressure in evaporation processes also causes high noise levels.

A substantial traffic load in the immediate vicinity of a dairy plant is generally unavoidable due to the regular delivery of milk (which may be on a 24-hour basis), deliveries of packaging and the regular shipment of products.

Noise problems should be taken into consideration when determining plant location. Hazardous wastes consist of oily sludge from gearboxes of moving machines, laboratory waste, cooling agents, oily paper filters, batteries, paint cans etc. At present, in western Europe some of these materials are collected by waste companies. While some waste is incinerated, much is simply dumped. Temperatures are often specified by regulation. Thermal energy, in the form of steam, is used for heating and cleaning. As well as depleting fossil fuel resources, the consumption of energy causes air pollution and greenhouse gas emissions, which have been linked to global warming.

Dairy products such as milk, cream and yogurt are typically packed in plastic-lined paperboard cartons, plastic bottles and cups, plastic bags or reusable glass bottles. Other products, such as butter and cheese, are wrapped in foil, plastic film or small plastic containers. Milk powders are commonly packaged in multi-layer kraft paper sacs or tinned steel cans, and some other products, such as condensed milks, are commonly packed in cans. Breakages and packaging mistakes cannot be totally avoided. Improperly packaged dairy product can often be returned for reprocessing; however the packaging material is generally discarded.

Emissions to air from dairy processing plants are caused by the high levels of energy consumption necessary for production. Steam, which is used for heat treatment processes (pasteurisation, sterilisation, drying etc.) is generally produced in on-site boilers, and electricity used for cooling and equipment operation is purchased from the grid.

Air pollutants, including oxides of nitrogen and sulphur and suspended particulate matter, are formed from the combustion of fossil fuels, which are used to produce both these energy sources.

In addition, discharges of milk powder from the exhausts of spray drying equipment can be deposited on surrounding surfaces. When wet these deposits become acidic and can, in extreme cases, cause corrosion.

For operations that use refrigeration systems based on chlorofluorocarbons (CFCs), the fugitive loss of these gases to the atmosphere is an important environmental consideration, since CFCs are recognised to be a cause of ozone depletion in the atmosphere. For such operations, the replacement of CFC-based systems with non-or reduced-CFC systems is thus an important issue.

Some processes, such as the production of dried casein, require the use of hammer mills to grind the product. The constant noise generated by this equipment has been known to be a nuisance in surrounding residential areas. The use of steam injection for heat treatment of milk and for the creation of reduced pressure in evaporation processes also causes high noise levels.

A substantial traffic load in the immediate vicinity of a dairy plant is generally unavoidable due to the regular delivery of milk (which may be on a 24-hour basis), deliveries of packaging and the regular shipment of products.

Noise problems should be taken into consideration when determining plant location. Hazardous wastes consist of oily sludge from gearboxes of moving machines, laboratory waste, cooling agents, oily paper filters, batteries, paint cans etc. At present, in western Europe some of these materials are collected by waste companies. While some waste is incinerated, much is simply dumped.

Water Consumption

As with most food processing operations, water is used extensively for cleaning and sanitising plant and equipment to maintain food hygiene standards. Due to the higher costs of water and effluent disposal that have now been imposed in some countries to reflect environmental costs, considerable reduction in water consumption has been achieved over the past few decades in the dairy processing industry. These improvements are attributed to developments in process control and cleaning practices.

At modern dairy processing plants, a water consumption rate of 1.3–2.5 litres water/kg of milk intake is typical, however 0.8–1.0 litres water/kg of milk intake is possible (Bylund, 1995). To achieve such low consumption requires not only advanced equipment, but also very good housekeeping and awareness among both employees and management.

Effluent Discharge

Dairy processing effluent contains predominantly milk and milk products which have been lost from the process, as well as detergents and acidic and caustic cleaning agents. The constituents present in dairy effluent are milk fat, protein, lactose and lactic acid, as well as sodium, potassium, calcium and chloride. Milk loss to the effluent stream can amount to 0.5–2.5% of the incoming milk, but can be as high as 3–4%.

A major contributing factor to a dairy plant's effluent load is the cumulative effect of minor and, on occasions, major losses of milk. These losses can occur, for example, when pipework is uncoupled during tank transfers or equipment is being rinsed.

The organic pollutant content of dairy effluent is commonly expressed as the 5-day biochemical oxygen demand (BOD5) or as chemical oxygen demand (COD). One litre of whole milk is equivalent to approximately 110,000 mg BOD5 or 210,000 mg COD.

Concentrations of COD in dairy processing effluents vary widely, from 180 to 23,000 mg/L. Low values are associated with milk receipt operations and high values reflect the presence of whey from the production of cheese. A typical COD concentration for effluent from a dairy plant is about 4000 mg/L. This implies that 4% of the milk solids received into the plant is lost to the effluent stream, given that the COD of whole milk is 210,000 mg/L and that effluent COD loads have been estimated to be approximately 8.4 kg/m3 milk intake.

A Danish survey found that effluent loads from dairy processing plants depend, to some extent, on the type of product being produced. The scale of the operation and whether a plant uses batch or continuous processes also have a major influence, particularly for cleaning. This is because small batch processes requires more frequent cleaning. The tendency within the industry towards larger plants is thus favourable in terms of pollutant loading per unit of production.

Water Consumption Survey for Danish Dairy Processing Plants

A survey of 72 Danish dairy companies operating a total of 134 processing plants was conducted in 1989 (Danish EPA, 1991). The product mix of the companies surveyed was as follows: 44 dairies produced butter, 90 produced cheese, 29 were market milk plants and 11 produced concentrates including milk powder. The plants surveyed were all technologically advanced and most claimed that they had reduced the pollutant load of their effluents by 30–50% compared with previous years. The survey found that on average each tonne of milk processed resulted in the production of 1.3 m3 of effluent with the following characteristics:

COD 2000 mg/L

BOD5 1500 mg/L

Fat 150 mg/L

Total nitrogen 100 mg/L

Total phosphorus 30 mg/L

Due to the traditional payment system for raw milk (which is based on the mass or volume delivered plus a separate price or premium for the weight of milk fat), the dairy processing industry has always tried to minimise loss of milk fat. In many countries the payment system now recognises the value of the non-fat milk components. Systems that control the loss of both fat and protein are now common in the industrialised world, but less so in the developing world. The disposal of whey produced during cheese production has always been a major problem in the dairy industry. Whey is the liquid remaining after the recovery of the curds formed by the action of enzymes on milk. It comprises 80–90% of the total volume of milk used in the cheese making process and contains more than half the solids from the original whole milk, including 20% of the protein and most of the lactose. It has a very high organic content, with a COD of approximately 60,000 mg/L (Morr, 1992). Only in the past two decades have technological advances made it economically possible to recover soluble proteins from cheese whey and, to some extent, to recover value from the lactose.

Most dairies are aware that fat and protein losses increase the organic load of the effluent stream and, even in the developing world, the use of grease traps has been common for some decades. Many companies, however, do not take any action to reduce the organic pollution from other milk components. It is becoming more common for dairy companies to be forced by legal or economic pressures to reduce the amount and concentration of pollutants in their effluent streams.

Therefore, at most sites, wastewater treatment or at least pretreatment is necessary to reduce the organic loading to a level that causes minimal environmental damage and does not constitute a health risk. The minimum pretreatment is usually neutralisation of pH, solids sedimentation and fat removal.

Energy Consumption

Energy is used at dairy processing plants for running electric motors on process equipment, for heating, evaporating and drying, for cooling and refrigeration, and for the generation of compressed air.

Approximately 80% of a plant's energy needs is met by the combustion of fossil fuel (gas, oil etc.) to generate steam and hot water for evaporative and heating processes. The remaining 20% or so is met by electricity for running electric motors, refrigeration and lighting.

The energy consumed depends on the range of products being produced. Processes which involve the concentration and drying of milk, whey or buttermilk for example, are very energy intensive. The production of market milk at the other extreme involves only some heat treatment and packaging, and therefore requires considerably less energy.

Energy consumption will also depend on the age and scale of a plant as well as the level of automation. To demonstrate this.

Table Energy Consumption for a Selection of Milk Plants

Type of Plant Total Energy Consumption

Plants producing powdered milk exhibit a wide range of energy efficiencies, depending on the type of evaporation and drying processes that are used. Energy consumption depends on the number of evaporation effects (the number of evaporation units that are used in series) and the efficiency of the powder dryer. Table 2–6 provides examples of how different evaporation and drying systems can affect the energy efficiency of the process.

Substantial increases in electricity use have resulted from the trend towards automated plant with associated pumping costs and larger evaporators as well as an increase in refrigeration requirements.

High consumption of electricity can also be due to the use of old motors, excessive lighting or possibly a lack of power factor correction.

Table Energy Consumption for Evaporation and Drying Systems

Type of evaporation and drying system Total energy consumption (GJ/tonne product)

5-effect evaporator and 2-stage drier 13–15

3-effect evaporator and 1-stage drier 22–28

2-effect evaporator and 1-stage drier 40–50.

Growth and Development

As a family of transcription factor, Myocyte Enhancer Factor-2 (MEF2) proteins are important regulators of cellular differentiation and consequently play a critical role in embryonic development (Potthoff and Olson, 2007).

The MEF2 gene family is widely expressed in all branches of eukaryotes from yeast to human. There are four subtypes of the MEF2 gene in vertebrates (human versions are denoted as MEF2A, MEF2B, MEF2C and MEF2D) rather than just one copy in Drosophila.

They are all expressed in distinct but overlapping patterns during embryogenesis till adulthood. All of the mammalian MEF2 gene family share two domains, MADS-box (56 amino acids) and MEF2 domain (29 amino acids). Their sequences were highly conserved in their N-terminal and diverge in C-terminal.

MEF2 family are essential for vertebrate skeletal muscle development and differentiation, which are selectively expressed in differentiated myocytes and activates nearly all skeletal and cardiac muscle genes by binding a conserved A/T-rich DNA sequence. During myogenesis, MEF2A and bHLH proteins cooperatively activate skeletal muscle genes and physically interact through the MADS domain of MEF2A and three myogenic amino acids of the muscle bHLH proteins

(Kaushal et al., 1994). Moreover, MEF2 controls skeletal muscle formation after terminal differentiation in nascent fibres and drives expression of genes encoding thick filament proteins.

Kaushal et al. (1994) proved that MEF2A induced myogenic development, when ectopically expressed in clones of nonmuscle cells of human clones.

The MEF2A gene previously had been linked to patients with coronary artery disease and myocardial infarction. In particular, a 21-bp deletion and missense mutations were demonstrated either to reduce MEF2A transcriptional activity or to impair its nuclear translocation. From 90-270 days, MEF2A expressions level were significant varied among cardiac muscle, dorsal muscle and leg muscle and the highest was leg muscle, the second was dorsal muscle, the lowest was cardiac muscle (Gao et al., 2009). Recently it was reported that the polymorphism of chicken MEF2A gene associated with body weight (Zhou et al., 2010).

The DNA sequencing and Forced-PCR-RFLP were used to identify the polymorphisms of the MEF2A gene in three Chinese indigenous cattle breeds. And the associations between the Single Nucleotide Polymorphisms (SNPs) within the bovine MEF2A gene and the growth traits of cattle were investigated, in order to identify the potential molecular markers in assistant cattle breeding.

Materials and Methods

Animals: Genomic DNA samples were obtained from 1009 individuals belonged to three cattle breeds: Qinchuan (QC, n = 287), Nanyang (NY, n = 272), Jiaxian (JX, n = 450). Growth traits of Nanyang and Jiaxian population were collected every six monthss from birth to twenty-four monthss (birth, 6, 12, 18, 24 months), including birth weight, Body Weight (BW), Hucklebone Width (HBW), Withers Height (WH), Heart Girth (HG), Body Length (BL) and Heart Girth Index (HGI).

Growth traits of Qinchuan cattle were collected after 3 years old including WH, BL, HBW, HG, BLI, HGI, Height at Hip Cross (HHC), Rump Length (RL) and Hip Width (HW).

PCR conditions: According to the whole genome shotgun sequence of Bos taurus MEF2A (GenBank accession No. NC_007319.3), nine pairs of primers were designed to investigate the potential SNPs in the MEF2A gene, The 15 ìL PCR solution contained 30 ng DNA templates, 1 ìmolL^{-1} of each primer, 2xReaction Mix 7.5 ìL (500 ìM dNTP each, 200 mmolL^{-1} Tris-HCl, 100 mmol L^{-1} KCl, 3 mmol L^{-1} MgCl$_2$) and 0.25 U Taq DNA polymerase (Tiangen, Beijing, China). DNA pool was mixed with 100 DNA samples with equal concentration (50 ng ìL^{-1}) for each group. And then the PCR products of DNA pool were used to sequence. The PCR was performed using the following program: an initial step at 94°C for 4 min, followed by 35 cycles (denaturation at 94°C for 40 sec, annealing temperatures for 40 sec and extension at 72°C for 40 sec) and there is a final extension at 72°C for 10 min.

Forced PCR restriction fragment length polymorphism (F-PCR-RFLP): Three mutations were detected using the ABI 377 sequencer from both directions (Applied Biosystems, USA) of pooling DNA samples after PCR amplification (NM_001083638: 1734 C>T (465aa), 1641 A>G (434aa), 1598 T>C (420aa)) in MEF2A gene. In order to exactly detect these mutations, the forced PCR-RFLP method was used. The primer sequences were in the following:

F1: *5'*-AGAGTTTGGGGGCCGGCCGAGCACACC-3'

R1: *5'*-AGGCCCCACAGCCGCAGCCCCGGCTGCA-3'

R2: *5'*-TCCAGCAGCAGCAGCAGCCACAGGCGC-3'

The underlined bases show the incorporated mismatch creating restriction sites. The length of PCR product was 191 bp, which contains three enzyme restriction sites to detect the three mutations of the MEF2A gene. In which 1734C>T, 1641A>G and 1598T>C could be detected by EcoR II, Pst I and BspT107 I, respectively.

PCR products were obtained from all individuals in this study and aliquots of 5 ìL PCR products were digested with 4U EcoR II, Pst I and BspT107 I (Takala, Dalian, China) for 10 h at 37°C, respectively.

The digested products were detected by electrophoresis in 3.0% agarose gel stained with ethidium bromide.

Statistical analysis: Genotypic and allelic frequencies of the 1598T>C, 1641A>G and 1734C>T were directly calculated. All sequences determined in this study were edited using the DNA star 5.0 package.

The associations between SNP marker genotypes and growth traits in cattle were analysed by the least-squares method as applied in the General Liner Models (GLM) procedure of SAS (SAS Institute Inc., Cary, NC, USA) according to the following linear model:

$$Y_{ijklmn} = \grave{i} + B_i + F_j + M_k + G_l + S_m + e_{ijklmn}$$

Where:

Y_{ijklm}	=	Observed value
$\grave{i}$	=	Overall mean for each trait
B_i	=	Fixed effect of ith breed
F_j	=	Fixed effect of jth farm
M_k	=	Fixed effect of kth months of surveying
G_l	=	Fixed effect of lth single SNP marker genotype
S_m	=	Fixed effect of sex
e_{ijklmn}	=	Random error

Results and Discussion

Nine exons of MEF2A gene were scanned in three Chinese indigenous cattle breeds and three mutations in exon 11 were identified. The locus C1598T, GenBank accession No. NM_001083638, same below) was identified as a missense mutation leading to a proline (CCG) to leucine (CTG) exchange. The loci G1641A and C1734T were all silent mutation.

The allele frequency and genotype frequency of the MEF2A gene were presented in. The C1598T and G1641A loci showed low Polymorphism (PIC = 0.13) and intermediate Polymorphism (PIC = 0.35) in NY but no polymorphism was found in QC and JX. In locus C1734T, the CC was the dominant genotype in NY and QC cattle but no polymorphism was detected in JX. The relationship between SNPs and the cattle growth traits were analysed. The individuals with genotype 1598CT and CC had larger BL and HG at 12 months than the individuals with genotype 1598TT ($p < 0.05$). This missense mutation maybe affects protein localization and transcription activity and then affects the cattle growth and development. The individuals with genotype 1641AA had better Average Daily Gain (ADG) at 12 months than the individuals with genotype 1641AT ($p < 0.05$).

The genotype 1734TT was better than the genotype 1734CT in BW and ADG at 6 months in cattle ($p < 0.05$). Although, the two silence mutations didn't cause amino acid change, which could affect the growth traits cattle through codon usage biases or mRNA structure changing, further experiments were needed to investigate the mechanisms. If $r^2 > 0.33$, then the linkage disequilibrium was considered strong (Li et al., 2009). The linkage disequilibrium between the three

SNPs were estimated, which indicated that the three SNPs were not linked strongly in the analysed populations (r^2 = 0.001, 0.023 and 0.006, respectively).

The MEF2A gene has highly conserved N-terminal MADS-box and MEF2 domains, while diverge C-terminal transactivation domain. No polymorphism was detected except exon11 in the nine exons of MEF2A gene in this study showed the consistent conservation mode.

The nucleotide sequence numbering shown is of the human MEF2A sequence and the percent sequence identities are all relative to Human MEF2A. The domain organization and sequence comparison of MEF2 proteins from representative species. The amino acid numbering shown is of the human MEF2A sequence and the percent sequence identities are all relative to hMEF2A The MEF2 proteins were involved in regulation of many muscle specific genes and played active roles in myogenesis, proliferation and differentiation. The functional importance of the MEF2A gene and the phenotype influence of the mutation implied that the three SNPs detected in this study could be potential markers in molecular marker assisted selection of cattle breeding. But only SNP association analysis is not enough, further research need to be done to investigate the mechanisms of these mutations and more research need to be done to investigate the MEF2 family's function in cattle's growth and development.

Growth and Development-Livestock and Poultry

MicroRNAs (miR) are small RNA molecules (~22 nucleotides) that are important regulators of numerous biological processes, including organ and tissue morphogenesis and function. In this capacity, most miR inhibit protein synthesis by binding to the 32-untranslated region of targeted mRNA species. Hundreds of genes can be regulated in this fashion. The objective of this experiment was to evaluate expression of miR in mammary tissue from Holstein cows at different developmental and functional stages. Tissues were obtained from: prepubertal heifers (6 mo) that were (1) intact, (2) ovariectomized, (3) intact + estrogen, (4) ovariectomized + estrogen; (5) from primiparous cows, 100-250 d of gestation; (6) from lactating cows, 14 d lactation; (7) from cows during the dry period, 40 d dry and 20 d prepartum. Total RNA was extracted from three or four animals at each stage and pooled to determine patterns of miR expression by hybridization to a microarray containing modified RNA targets complementary to all known mirR.

Expression of miR such as miR-221 and miR-127 appeared to be differentially expressed prepubertally. Expression of miR-615 was enhanced by estrogen treatment and miR-29a by ovariectomy. During first gestation, expression of miR-20a was increased. During lactation, miR were typically expressed at low levels, but there was increased expression of a limited number of miR, including miR-326 and miR-350. During the dry period, there was increased expression of miR-542-5p and miR-690.

We subjected individual RNA samples to quantitative RT-PCR and confirmed patterns of expression revealed by microarray in 4 of 5 genes tested. Our quantitative RT-PCR results confirmed the utility of evaluating miR expression by microarray and suggested that miR function as regulators of mammary gland development and function. Ribonuclease protection assays revealed that injection of recombinant bovine GH in a slow-release formulaincreased both hepatic GHR and insulin-like growth factor I (IGF-I) mRNAs one week after initiation of treatment.

The increases in GHR and IGF-I mRNAs were highly correlated. Western blot analysisshowed that the injection also increased GHR protein level in the liver. In cattle and several other mammals, hepatic GHR mRNA is expressed as variants that differ in the 52-untranslated region, due to use of different promoters in transcription and/or alternative splicing. We found that GH injection increased the expression of the liver-specific GHR mRNA variant 1A (GHR1A), without affecting GHR1B andGHR1C mRNAs, the other two major GHR mRNA variants in the bovine liver.

Transient transfection analyses of a 2.7 kb GHR1A promoter in reconstituted GH-responsive cells showed that GHcould robustly activate reporter gene expression from this promoter, suggesting that GH augmentation of GHR1A mRNA expression in the liver is at least partially mediated at the transcriptional level. Further transfection analyses of serially 52-truncated fragments of this GHR1A promoter narrowed the GH-responsive sequence element down to a 210 bp region that contained a putative signal transducer and activator of transcription 5 (STAT5) binding site.

Electrophoretic mobility shift assays demonstrated that this putative STAT5 binding site was able to bind to STAT5b protein. In transfection assays, deletion of this putative STAT5 binding site abolished most of the GH response of the GHR1A promoter. These observations together suggest that GH stimulates the expression of one GHR mRNA variant, GHR1A, through binding STAT5 to its promoter, thereby increasing GHR protein expression in the bovine liver.

Temporal Longissimus Muscle gene Expression Profiles due to Plane of Dietary Energy in Early-weaned Angus Steers

Energy-dense nutrients might trigger long-term genomic adaptations of economic importance in skeletal muscle of young steer calves. Objectives were to evaluate temporal gene expression profiles in longissimus muscle (LM) of early-weaned (~140 d age) Angus steers (n = 6/diet) fed a high-grain (HiE, NE = 1.43 Mcal/kg) or high-by-product (HiF, NE = 1.19 Mcal/kg) diet for 120 d, at which point all steers were switched to a common feedlot diet until slaughter. LM biopsies for transcript profiling and blood for metabolite analyses were collectedat 0, 60, and 120 d of feeding. BW, ADG, back fat (d 60 and 120), and marbling scores (d 60 and 120) also were measured. A 13,257 bovine oligonucleotide (70-mers) array was used for transcript profiling.

Annotation was based on similarity searches using BLASTN and TBLASTX against human, mouse, and bovine UniGene databases, the human genome, and the cattle TIGR database. Cy3-and Cy5-labelledcDNA from LM and a reference standard were used for hybridizations. Feeding HiE vs. HiF resulted in greater (time × diet $P < 0.05$) temporal blood glucose concentrations (88 vs. 80 mg/dL on d 120), whereas HiF increased (time × treatment $P = 0.06$) blood â-hydroxybutyrate (BHBA) concentration (0.48 vs. 0.36 mmol/L on d 120) to a greater extent than HiE. ADG over the 120 d tended ($P = 0.08$) to be greater with HiE (3.6 vs. 3.4 kg/d). ANOVA (FDR $P = 0.10$) identified 504,67, and 141 differentially expressed genes due to time, diet, and diet ×time, respectively.

Genes associated with aspects of metabolism (e.g. protein or fatty acid synthesis), development, and signal transduction activity predominated among those affected by time × treatment. Results suggest that high plane of dietary energy during the early growth phase might improve efficiency of gain at least in part through the provision of a specific pattern of nutrients (e.g. glucose vs. BHBA) to skeletal muscle, which can in turn directly or indirectly promote genome-wide alterations in gene expression affecting tissue growth and development.

Creation of a Gene Atlas in Cattle using Sequence-based Transcriptional Profiling

Numerous opportunities to advance the understanding of how heritable variation affects economically important traits are being provided through resources generated from the Bovine Genome Sequencing project. Success in these investigations relies upon the depth in which the sequence assembly is annotated. In humans and biomedical model species, extensive annotation to identify genes and report relative levels of expression in various tissues has been accomplished by creation of Gene Atlas databases that provide researchers instant access to the expression profile of a gene under study. Similarly, we are constructing a Bovine Gene Atlas database that will house transcript profiles from 100 different bovine tissues collected from major organ systems. RNA was extracted using tissues derived from the genome sequencing cow and her offspring.

Transcript profiles were captured using a digital, sequence-based approach known as Sequence-By-Synthesis from a Clonal Single Molecule Array. Yield of 20 bp cDNA sequence tags exceeded more than 5 million counts per sample. The more than 500 million tags were grouped according to tissue of origin, sequence similarity and genome map position in order to assign gene identities and account for potential sequencing errors.

To test the correlation between the sequence tag data and a known pathway for synthesis of 3, 16 or 17-Glucuronide, tag counts for gene members of this pathway were compared between adult testes, two stages of uterus development, muscle, and placentome. Tag count analysis revealed this metabolic pathway is greater than 50 fold more active in ovary and uterusversus testes, placentome, and muscle. We conclude relative levels of expression for nearly all genes, even those for rare and species-specific transcripts, can be accurately determined. Such a framework of expression data will allow determination of regional transcriptional control, tissue phylogeny and interconnected gene networks.

Backgrounding-Feeder Cattle Nutrition

The primary objective of backgrounding is to provide optimal growth and development of the muscle and frame of the calf, while avoiding excess fat deposition. After making an inventory of resources including feed, labour and cash reserves, a partial budget may indicate a net advantage to place calves on feed.

Objectives and Opportunities

The primary objective of backgrounding is to ensure optimal development and growth of the muscle and frame, while avoiding excess fat deposition. After taking an inventory of resources, including feed, labour and cash reserves, develop a partial budget. This will help to determine if there is a net advantage to placing calves on feed.

Analyse Marketing Opportunities

Investigate marketing options with cattle buyers or feedlot operators to determine the market outlook and expected sale prices before making a final decision to background calves.

In order to raise a healthy, 800 to 900-lb. feeder, the producer must consider the animal's frame size and intended market when selecting a feeding program. Frame size affects growth and development. Large-frame calves grow faster and as such, have greater daily dry matter intakes than small and medium-frame calves. They also fatten and finish at heavier weights.

Other factors — such as fleshiness, quality, sex and breed — must also be considered. The backgrounded feeder should be "green," not carrying excess fat, and ready to be placed on a high energy finishing ration in a feedlot. Some smaller-framed calves may be placed on a summer grassing program the following year.

Most backgrounding rations are designed to result in a predetermined daily rate of gain. It is important to feed test and to use the services of livestock nutritionists when developing backgrounding feeding programs.

Target Weights and Target Marketing Date

Once the frame size of the feeder calves has been determined, three fundamental factors will dictate the design of the feeding program:

1. Target Sale Weight
2. Target Marketing Date
3. Condition or Degree of Fleshiness.

Target Sale Weight

The target sale weight and average daily gain varies according to frame size. Producers need to determine the type of calves they will feed. It is important to realize that target sale weights will be affected by the price of feed and current feeder markets. For example, when fed cattle prices decline, feedlots become more sensitive to feed grain prices. Large-frame calves can be placed directly onto a finishing program after weaning with minimal backgrounding. Some feedlots, however, may sell finished large-frame cattle to specific Canadian or American markets which target cattle with heavier weights and a higher degree of marbling.

These feedlots may be interested in purchasing backgrounded large-frame cattle. Before starting a backgrounding feeding program using large-frame calves, be sure you have a market for them.

Target Marketing Date

Knowing the marketing date and sale weight allows the producer to calculate the number of days a feeder animal will be on feed and the total amount of gain required to reach the market weight.

A ration can then be formulated, based on feed test results, to produce the daily gain required to meet the objectives of the feeding program.

For example:

- Buy or wean a medium-frame 500 lb. calf on November 15
- Target marketing date is April 15
- Days on feed = 150 days
- Target sale weight is 800 lb.
- Total weight gain = 300 lb.
- 300 lb. gain/150 days = 2.0 lb. gain per day *

*(Average Daily Gain for the entire feeding period)

A nutritional program can now be designed to provide a ration that will provide 2.0 lb. of average daily gain.

Six Steps in Formulating a Ration;

1. Determine nutrient requirements, feed intake and desired weight gain for each class of cattle.
2. Feed-test "on-farm" feeds to determine nutrient levels.
3. Determine required "off-farm" feedstuffs (protein supplements, minerals, feed additives, vitamins, etc.)
4. Formulate the rations.
5. Implement the nutritional program and monitor the performance of the cattle.
6. Adjust rations according to weather conditions and animal performance.

The National Research Council (NRC) publishes guidelines for beef cattle nutrition. These prediction equations have been further modified by applied research conducted in Western Canada in order to meet the nutritional requirements for feeding cattle in Saskatchewan. These guidelines have been used to formulate the ration examples in this fact sheet.

21-Day Weaning Period

If the calves have not been previously weaned, place them on a 21-day weaning program prior to backgrounding. This helps the calves through the stressful weaning period and encourages them to eat grain and long hay or silage. A good supply of accessible clean fresh water is essential.

Feeding the calves two or three times per day during the weaning period will help them become accustomed to eating dry feeds. It also acquaints them with regular handling and provides an opportunity to observe behaviour and identify sick animals. Avoid feeding on the ground. This practice results in considerable feed wastage and allows for transmission of disease.

Some producers prefer feeding good quality long hay for the first week. Avoid the sudden introduction of alfalfa hay as it may cause scouring or bloat. Some instances of bloat have occurred when alfalfa grass hay was fed. Calves will sort the hay and may selectively eat the alfalfa portion. Avoid over-processing or over-grinding if using a bale shredder or grinder.

Oats, barley or fortified pellets may be offered in addition to the free-choice hay. Start the calves by feeding one pound of concentrate per head per day until each calf is consuming the concentrate. Increase the concentrate by one-half pound per head per day until the desired level is reached. Watch for signs of digestive upset such as a reduction in feed intake, bloating or scours. Other options include silage mixed with barley or oats, supplements, minerals and vitamins. Calves may refuse to eat large amounts of silage. The amount of silage fed should be gradually increased. Straw should be available on a free-choice basis.

Backgrounding Period

Many backgrounding rations contain 60 to 70 per cent forage (dry matter basis), with the balance comprised of grain or fortified pelleted grain screenings. As the backgrounded calves mature, the energy component or Total Digestible Nutrients (TDN) of the rations is gradually raised by increasing the amount of grain or pellets fed.

Most backgrounding rations require additional salt and minerals. Trace Mineralized Fortified Salt (TM Fortified Salt) is recommended. In addition to cobalt and iodine, it contains a number of required trace minerals (copper, zinc, manganese and sometimes selenium). Research has demonstrated that these trace minerals are commonly deficient in Saskatchewan-grown forages and grains used for beef cattle production.

Calcium and phosphorus are important for proper skeletal growth and development in backgrounded calves. If the forage and grain component of the rations does not supply adequate levels of these minerals, they must be provided in a mineral supplement.

Feed Additives and Implanted Growth Promotants

There are a number of feed additives that can improve the health and productivity of beef cattle. Feed additives can be grouped into six categories: ionophores, synthetic hormones, antibiotics, probiotics, coccidiostats and bloat prevention aids.

Ionophores are compounds that alter the rumen microflora to increase the production of propionic acid. The net effect is improved feed efficiency (less feed is required to maintain normal or improved rates of gain). Some ionophores may increase the average daily rates of gain. Ionophores may reduce the incidence of grain overload and bloat, and some act as "anticoccidial" agents.

Some products contain synthetic hormones which are designed to suppress estrus (heat) in beef heifers intended for slaughter. Feed efficiency and average daily gains may be improved. Registered uses of antibiotics include treatment of stress-related sickness and, in some cases, growth promotion. Some antibiotics aid in the prevention of bloat, foot rot and diarrhea. Probiotics favour the growth and development of the more desirable rumen microbes. They may improve overall performance.

Coccidiostats are used to prevent "coccidiosis," a disease caused by parasitic protozoa. Symptoms of this disease are bloody scours and loss of performance. Coccidiosis is more prevalent in young cattle raised in confined conditions.

Growth promotant's can increase daily rates of gain by eight to 10 per cent and improve feed efficiency. A number of products are currently registered by the federal government for use in beef cattle. They are classed as natural steroid, synthetic steroid-like and non-steroid compounds. Growth promotant's are implanted or deposited under the skin in the middle third of the ear. Implants may be combined with some feed additives to further increase performance.

Always follow label directions and observe pre-slaughter withdrawal periods when using feed additives or implants. A veterinarian or livestock nutritionist should be consulted before using any of these products in a feeding program.

Preventing Grain Overload

Calves that consume excessive amounts of grain or pellets may develop grain overload. In mild cases, symptoms include reduced feed intake. In more severe cases, signs include abdominal pain (calves kick at their bellies), bloating, sunken eyes (dehydration) and diarrhea.

Light coloured feces may indicate digestive problems. Some calves may develop laminitis or founder. Severely bloated calves may die if not treated. Under normal conditions, the microbes in the rumen produce a number of compounds, including acids, as the starches in the grain and pellets are fermented. These acids provide energy to the growing calf. If an excessive amount of starch is available to the microbes, high amounts of acids and bacterial slime are produced. This combination can lead to feedlot bloat and grain overload.

To reduce or prevent grain overload, process grain by cracking each kernel into two or three pieces. This ensures that the particle size is not too small. Mixing the grain or pellets with coarsely ground forage or silage ensures that the roughage is consumed with the grain or pellets. This helps to prevent grain overload.

Another option would be to feed smaller amounts of grain or pellets several times during the day. Feeding 10 lb. of concentrate to a 500 lb. calf at one feeding may cause grain overload, especially if the calf was hungry. Splitting the concentrate into several three or four-pound meals spread out over the course of the day should eliminate digestive problems.

Feedbunk Management

Good feedbunk management is important to keeping calves on feed and minimizing digestive disturbances such as grain overload. The goal is to consistently deliver the proper amount of feed at the right time. Ensure that there is adequate space at the feedbunk so that all calves can eat at the same time. Over-consumption occurs when calves are hungry.

This increases the incidence of digestive upsets. Changes in weather will also affect feed consumption. Calves tend to eat more during stormy weather. Although research indicates differing opinions, feeding twice per day often improves feed efficiency. It also ensures that the feed is fresh and reduces problems if mixing errors occurred. Keep feedbunks free of old feed and manure. They should be cleaned weekly.

Buying vs. Raising Replacement Heifers

When deciding on the best strategy for replacing heifers, producers need to weigh the advantages and disadvantages of raising or buying replacement females as well as consider other economic and general management issues specific to their operations. Factors to consider include:

- Current and future market prices
- Herd size
- Pastures, facilities and management level
- Available labour
- Economics
- Herd health concerns
- Cow genetic base (crossbreeding system)
- Herd quality
- Purchase replacement alternatives.

To clarify which strategy is best for a specific operation, producers should develop individualized budgets and management plans for each option.

Current and Future Market Prices

The beef industry is cyclical, with a series of high and low prices occurring about every 10 years. The law of supply and demand governs these cycles. As in other businesses, when supplies are down and demand is steady, prices tend to rise.

When cattle prices are high, producers begin to rebuild their herds by retaining "high value" heifers or by purchasing replacements. The thinking is that with high cattle prices, it is time to get into beef production or to increase current cow inventories. After the rebuilding phase occurs, supplies increase and prices drop. This is the beginning of the herd liquidation phase of the cattle cycle.

Another explanation of the cattle cycle is that cash flow often determines the number of heifers retained or purchased. When prices are low, producers often must sell more or buy fewer heifers to meet cash flow demands. Conversely, as prices rise, producers are able to sell fewer heifers to meet cash flow demands. Thus, a common joke in the beef industry is "buy high and sell low."

Buying or retaining more replacements when prices are high is contrary to good business principles. Another problem with this practice is that heifers born during periods of high prices will produce calves during the following period of low prices, and vice versa.

To improve cow-calf profitability, producers need to adjust their replacement strategies. A study of replacement strategies by Iowa State University in 2001 examined production and financial data from 1970 to 1999. The strategies that were studied included:

- Maintaining the same number (SS) of heifers each year
- Maintaining the same cash flow (CF) each year—when calf prices are high, the producer retains or buys more heifers
- Retaining the same dollar value (DV) of heifers each year— when calf prices are low, the producer retains more heifers.

The researchers found that the return over cash costs for the DV strategy was 55 percent higher than the CF strategy and 33 percent higher than the SS strategy. These findings indicate that it is more profitable to use countercyclical replacement strategies. That is, they should purchase more replacements when calf prices are low. However, producers using a countercyclical strategy must be able to weather large variations in cash flow.

Cycles are affected by changes in consumer demand, environmental conditions that affect production, and other unforeseeable events that can affect the market, such as the cases of bovine spongiform encephalopathy (BSE, or mad cow disease) in Canada and United States. To make informed decisions, the producer must evaluate the current market situation and develop an individualized budget.

Bibliography

Abdullah, A Mohamed : *Food Security and Gender Inequality*, Abjiheet, Delhi, 2008.

Alec, William: *The Dairy Chemical Industry*, London: Longman Group Limited, 1971.

Arora, Dinesh: *Biotech's Dictionary of Dairy Science*, Biotech Books, Delhi, 296.

Avanish K. Tiwari: *Food Security and Global Economy*, Pentagon Press, Delhi, 2009.

Babita Bohra: *Dairy Farming in Mountain Areas*, Daya, Delhi, 2006.

Basavaraj S. Benni, Rawat: *Dairy Co-operative Management and Practice*, Delhi, 2005.

Bhattacharya, Lata: *Biochemistry of Nutrition*, Discovery, Delhi, 2010.

Bhutani, R.C.: *Fruit and Vegetable Preservation*, Biotech Books, Delhi, 2003.

Bohra, Babita: *Dairy Farming in Mountain Areas*, Daya, Delhi, 2006.

Brij K. Taimni: *Food Security in 21 Century : Perspective and Vision*, Konark, Delhi, 2001.

Brock, H. : *History of Dairy Chemistry*, New York: Norton, 1992.

Bucciarelli, L.: *Designing Engineers in Dairy* , Cambridge: MIT Press, 1995.

Bushnell, R.B. : *Dry Cow Feeding and Management*, A Western Regional Extension Publication, 1979.

Chakraborty, Sudip : *Food Security and Child Labour : The Case of a Hazardous Occupation*, Deep and Deep, Delhi, 2011.

Chand, Ram: *Decision-Making of Dairy Beneficiaries: Role of Aspiration, Motivation and Knowledge*, Om Pub, Delhi, 2010.

Chaturvedi, Pradeep : *Food Security and Panchayati Raj*, Concept, Delhi, 1997.

Chhazllani, V K : *Dairy Chemistry and Animal Nutrition*, Manglam Pub, Delhi, 2008.

Cohen, Lizabeth, *Making a New Deal, Industrial Workers in Chicago, 1919-1939* Cambridge University Press, 1991.

Collymore L.: *Fruit Production in Barbados*, Port of Spain, Trinidad and Tobago, 1996.

Cristobal Noe Aguilar : *Food Science and Food Biotechnology in Developing Countries*, Asiatech Pub, 2008.

Damasio, AR.: *Descarte's Error: Emotion, Reason and the Human Brain*, New York, 1994.

David, M.: *Ideas in Chemistry: A History of the Science*, New Brunswick, N.J.: Rutgers University Press, 1992.

De, Sukumar: *Outlines of Dairy Technology*, Oxford University Press, Delhi, 2001.

Devraj B: *Impact of Dairy On Small and Marginal Farmers*, Prateeksha Publications, Delhi, 2010.

Dinesh Arora: *Biotech's Dictionary of Dairy Science*, Biotech Books, Delhi, 296.

Droop, H. Richmond: *Laboratory Manual of Dairy Analysis,* Biotech, 2004.

Fox, Patrick F.: *Advanced Dairy Chemistry: Proteins*, New York: Elsevier Applied Science, 1992.

Friberg, Stig. E..: *Food Emulsions*, New York: M. Dekker, 1997.

Guarti, Luigi, *The Valuation of Firms*, Blackwell Publishing, 1994.

Gunjan Goel : *Applied Dairy and Food Microbiology,* Agrotech, 2005.

Hui, Yiu H. : *Dairy Science and Technology Handbook*, New York: Wiley, 1993.

Jacobson, M. : *Safe Food: Eating Wisely in a Risky World,* Washington, DC: Living Planet Press, 1991.

Jensen, Robert G. : *Handbook of Milk Composition*, San Diego: Academic Press, 1995.

Jha, S N : *Dairy and Food Processing Plant Maintenance : Theory and Practice,* International Book Distributing, Delhi, 2006.

John Prince: *Dairy Farming: Being the Theory, Practice, and Methods of Dairying*, New York, 1888.

Johnston, J. R. : *Molecular Genetics of Yeast, a Practical Approach.* IRL Press, Oxford 1994.

Kango, Mangala: *Normal Nutrition: Fundamental and Management*, RBSA, Delhi, 2003.

Kapoor, Ajay : *Dairy Science and Technology*, Vishvabharti Pub, Delhi, 2005.

Koli, P. A.: *Dairy Development in India: Challenges Before Co-Operatives*, Shruti Pub, Delhi, 2007.

Krimsky, Sheldon : *Biotechnics and Society: The Rise of Industrial Genetics*, New York: Praeger Publishers, 1991.

Kumar, Shashi : *Biodiversity and Food Security*, Atlantic, Delhi, 2002.

Law, Barry A.: *Microbiology and Biochemistry of Cheese and Fermented Milk*, London: Blackie Academic & Professional, 1997.

Leena Parihar: *Dairy Microbiology*, Agrobios, Delhi, 2008.

Michael, J. Lewis,: *SeparationProcesses in the Food and Biotechnology Industries*, Cambridge, U.K.: Woodhead, 1996.

Modi, H.A. : *Dairy Microbiology*, Aavishkar Publishers, Delhi, 2009.

Mohana Swamy : *Agro's Dictionary of Dairy Science,* Agro Botanical, 1995.

Mowery, D. C. : *Technology and Wealth of Nations,* Stanford: Stanford University Press, 1992.

Mudgal, V. D.; K. K. Singhal and D. D. Sharma: *Advances in Dairy Animal Production,* International Book Distribut, 2003.

Narang, R.K.: *Fruit and Vegetable Preservation Techniques*, APH Pub, Delhi, 2010.

Pandey, D. N. and Amita Bajpai: *Recent Trends in Animal Nutrition and Feed Technology for Livestock, Pets and Laboratory Animals*, International, 2003.

Parihar, Pradeep and Leena Parihar: *Dairy Microbiology*, Agrobios, Delhi, 2008.

Patton, Stuart: *Principles of Dairy Chemistry*, Huntington, N.Y.: Krieger, 1976.

Paul B. : *Food Biotechnology in Ethical Perspective*, Aspen, CO: Aspen Publishers, 1997.

Pirtle, Thomas Ross: *History of the Dairy Industry,* Chicago: Mojonnier Bros. Company, 1926.

Pradeep Chaturvedi: *Food Security in South Asia,* Concept, Delhi, 2002.

Qystein V. Sjaastad: *Physiology of Domestic Animals*, International Book Distributing Co., Delhi, 2005.

Ram Chand: *Decision-Making of Dairy Beneficiaries : Role of Aspiration, Motivation and Knowledge*, Om Pub, Delhi, 2010.

Ramakant Sharma : *Chemical and Microbiological Analysis of Milk and Milk Products,* International Book Distributing, 2006.

Rao, M K : *Food and Dairy Microbiology*, Manglam Pub, Delhi, 2007.

Rao, P. Venkateshwara : *Dairy Farm Business Management*, Biotech Books, Delhi, 2008.

Saini, M.L. : *Plant Breeding and Crop Improvement*, CBS, Delhi, 1997.

Samvel, A. P. V.: *Agri-Business Management*, Satish Serial Pub, Delhi, 2008.

Sarkar, A: *Advanced Organic Chemistry: Reactions and Mechanisms*, Swastik Publications, Delhi, 2011.

Shepherd, D. : *Homeopathy for the First Aider,* Sussex, England: Health Science Press, 1953.

Shukla , Arvind N.: *Textbook of Dairy Chemistry*, Discovery Pub, Delhi 2010.

Singh, Gajendra : *Food for All : An Assessment of Food Security in Indian Context*, MD Pub, Delhi, 2007.

Singh, Harmeet: *Dairy Farming*, APH, Delhi, 2005.

Singh, S K : *Biotechnology, Plant Propagation and Plant Breeding*, Campus Books, Delhi, 2008.

Singh, Vir and Babita Bohra: *Dairy Farming in Mountain Areas*, Daya, Delhi, 2006.

Singhal, K K and D D Sharma : *Advances in Dairy Animal Production,* International Book Distribut, 2003.

Spreer, Edgar : *Milk and Dairy Product Technology*, Translated by Axel Mixa. New York: M. Dekker, 1998.

Sriram Sridhar : *Enzyme and Food Biotechnology*, Wisdom Press, 2011.

Sujata K. Dass: *Biotechnology and Food Security*, Isha Books, Delhi, 2004.

Sukumar De: *Outlines of Dairy Technology*, Oxford University Press, Delhi, 2001.

Suri, Nitin : *Molecular Biology and Biochemistry*, Oxford Book Company, Delhi, 2010.

Susanna Hornig : *A Grain of Truth: The Media, the Public, and Biotechnology*, Lantham, MD: Rowman and Littlefield, 2001.

Thompson, Paul B. : *Food Biotechnology in Ethical Perspective*, Aspen, CO: Aspen Publishers, 1997.

Thomson, Sutherland: *Grading Dairy Produce*, Medi World Press, Delhi, 1995.

Tomar, S.K. and Gunjan Goel : *Applied Dairy and Food Microbiology,* Agrotech, 2005.

Tramontano, A., : *Antibodies as enzymes*, Trends in Biochemical Sciences,1987.

Tripathy, S.N. : *Food Biotechnology*, Dominant, 2004.

Tyagi, Prasum: *A Textbook of Animal Physiology*, Dominant, Delhi, 2010.

Upadhyay, K.G. and Vyas, S.H.: *Composition of Camel's Milk*, Gujarat Agric. University, 1982.

Venkateshwara Rao: *Dairy Farm Business Management*, Biotech Books, Delhi, 2008.

Walstra, Pieter, and Robert Jenness: *Dairy Chemistry and Physics*, New York: Wiley, 1984.

William Alec : *The Dairy Chemical Industry*, London: Longman Group Limited, 1971.

Yegge, Wilbur M., *A Basic Guide for Valuing a Company*, New York, Wiley, 1996.

Index

□□□